W9-BPJ-252

SAWBONES

the HORRIFYING, HILARIOUS ROAD *to* MODERN MEDICINE

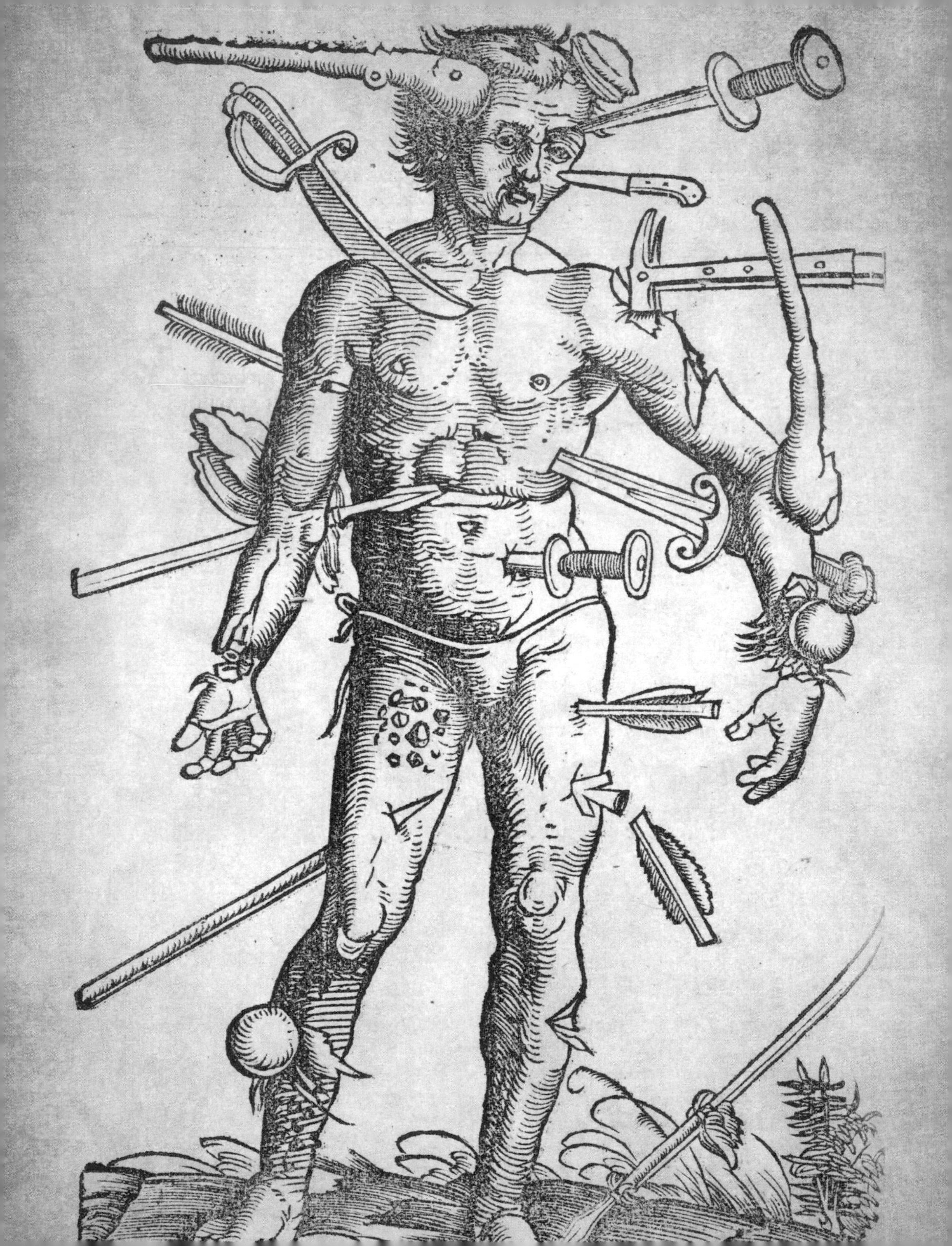

SAWBONES

the HORRIFYING, HILARIOUS ROAD *to* MODERN MEDICINE

Written by Dr. Sydnee McElroy
and Justin McElroy

Illustrated by Teylor Smirl

weldon**owen**

TABLE OF CONTENTS

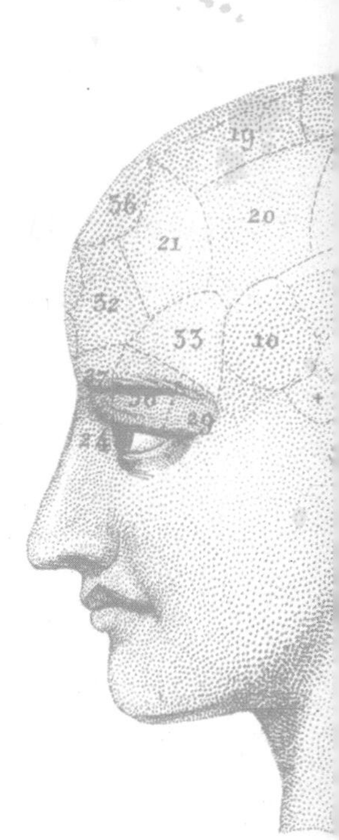

Part II: The Gross

Part III: The Weird

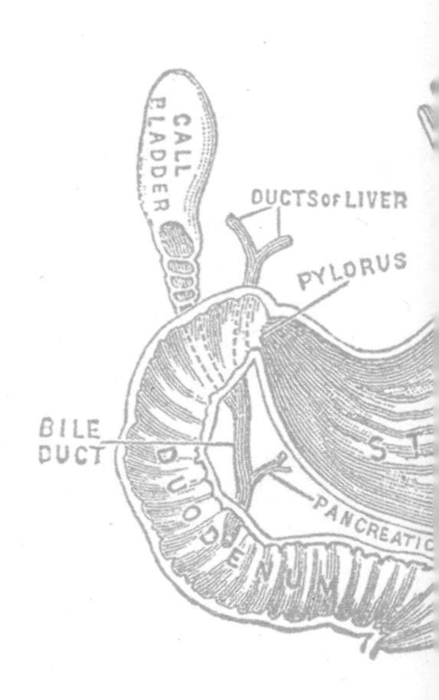

Part IV: THE AWESOME

This is a book about medical history and nothing we say should be taken as medical advice or opinion. It's for fun. Can't you just have fun for once and not try to diagnose your mystery boil? We think you've earned it. Just sit back, relax, and let this book distract you from that . . . weird growth. You're worth it.

For Charlie and Cooper,
thanks for sleeping sometimes
so we could write this.

We started Sawbones so we could stop watching so much TV.

After Losing the Sheen (a *Two and a Half Men* review show by two people who didn't watch until Charlie Sheen left the show) and Satellite Dish (a much more general and dare-we-say palatable TV review podcast) we just couldn't keep up with all the requisite screen time.

Sydnee has loved weird medical history since before she was a physician and Justin loves talking to Sydnee. A show that detailed humanity's absolutely ramshackle history of trying to fix itself over the millennia just made sense. One 45-minute brainstorming session at Black Sheep Burritos and Brews later, Sawbones was born.

We've never been the same.

Since we launched it in 2013, Sawbones has provided us with a surrogate family of medical history nerds who are just as passionate about science education as we are. After our daughter Charlie's frightening birth experience (she's fine now!) Sawbones was the place we wanted to share our story. We've received so many beautiful cards and letters from a new generation of physicians and scientists it has in some small way inspired. It's way more than a podcast to us.

We mean that literally too. We always thought Sawbones had the potential to reach new people with a book, we just had no idea how to make it happen. Considering that we treat podcasting as a family business, it should have come as no surprise that Sydnee's sister and incredibly gifted illustrator Teylor Smirl would provide a big part of the answer.

If you are a listener, the tales that follow may seem familiar. We started with some of our favorite episodes and dove deeper, expanding them into beautifully illustrated stories. Medical history has no shortage of ridiculous characters, misguided diagnoses, stomach-churning treatments and, occasionally, incredible miracles. Join us as we chronicle medicine's tortuous journey from complete ignorance to, well, something slightly more competent.

If you one of those listener who've been braving our horrible tales since 2013: Thank you, this book would not exist without you. If you're a newcomer, blame those other guys for what you're about to endure.

—Justin and Sydnee McElroy

THE UNNERVING

There are dark corners of existence that some people spend their lives trying not to think about. Let's start the fun there!

Before things get too wretched, let's unsettle and squirm
We'll steal bodies for science before they're too firm
You'll be shocked back to life after being starved dead
We've got poppies to take the edge off the dread
Stuff these herbs in your nose, the plague tends to stink
And then let's meet Pliny for an herb-and-pee drink
If higher awareness you're hoping to find
Don't puncture your noggin, we'll open your mind.

The Resurrection Men

The first riot in American history happened in Manhattan, and it happened because of dead bodies.

In the book biz, we call that a "tease." We'll get there, we promise, but first you need to understand the perfectly respectable and not in any way creepy reasons that doctors are so desperate to cut up dead bodies—not to mention how people over history have felt about that completely reasonable desire.

Autopsies haven't always been so controversial. In fact, the ancient Egyptians performed what were essentially autopsies way back in 2600 BCE, although the practice at the time was more about the ritualistic entombing of organs than any kind of education or forensics. Egyptian embalmers were the original anatomists, carefully removing a variety of body parts to be preserved while leaving the heart, eyes, and tongue in place for religious reasons. The challenges of extracting all those organs (especially the really slippery ones) led to the development of better surgical tools. Physicians of the time benefitted from those tools as well as learning a fair amount of anatomy from the embalmers, as demonstrated in ancient writings such as the Ebers Papyrus, the Edwin-Smith Papyrus, and the Kahun Gynecological Papyrus.

Everyone, I'd like you to meet my wife, who I suspect may be the only human on Earth that's this excited about these specific medical papyri. Admittedly, the Kahun Gynecological Papyrus is a very excellent name and worthy of highlighting. Gentle ribbing withdrawn.

Before you get too impressed, keep in mind these docs also performed pregnancy tests by putting an onion in the patient's vagina overnight, and thought that our arteries carried semen. So things weren't, you know, all sewn up, as it were. But still, an impressive start!

CLASSICAL CUT-UPS

In Ancient Greece, bodies were dissected without ritual, for purely scientific reasons. Erasistratus and Herophilus, who lived around 300 BCE, are known today as the fathers of modern dissection. They and a number of other physicians and students dissected corpses regularly and published texts and drawings based on their findings. (This was before printing presses, so let's spare a thought for the poor scribe who spent many icky afternoons copying their work.)

The practice wasn't considered legal in the strictest sense at the time, but it was apparently tolerated, and thus provided the Greeks with a better understanding of anatomy as well as an appreciation for the importance of autopsies in medical education.

In comparison, the Ancient Romans were strictly anti-autopsy—and by the time of Galen, the most prominent physician alive during the first century CE, dissections were made illegal. This meant that Galen and others had to base their know-how on primate anatomy . . . and the work of their Grecian predecessors. (If that last bit seems hypocritical to you, we have to assume this is the first account of human history you've ever read and we're so flattered that you've chosen our book to start with.) This reliance on secondhand information led to inevitable missteps and discord among physicians of the time, and for a millennium to follow.

Let this be a lesson kids: Always dissect your own corpses. You may think you can save a few bucks by looking over your buddy's shoulder, but trust me on this one. You've gotta get your own scalpel in there and saw through the sinew yourse—you know what? Actually, I've yucked myself out.

CAN I HAVE THAT WHEN YOU'RE DONE WITH IT?

After a millennium or so of European physicians and scholars poking around with carcasses and arguing about old scrolls, things started to pick up right around the 13th century. History tells us that dissections were not only carried out by doctors, but actually condoned by the Catholic Church. That may seem weirdly progressive, but keep in mind the doctrine that the body exists only as a vessel for the soul—and once that soul vacates the premises, why shouldn't scientists get a crack at the abandoned home?

AUTOPSIES FOR FUN AND PROFIT

At this point, dear reader, we're sure you're getting excited to learn what corpse dissection was like during the never-ending Insane Clown Posse concert that was Middle Ages. We're thrilled to report The Rowdy Years did not disappoint, since that's when autopsies became a spectator sport. Literally. As in, they were conducted in public and tickets were sold. Oh Middle Ages, thank you as always.

Dissections continued as the Middle Ages waned, though they tended to be more private ... even occasionally secretive. For example, in the 15th century, Leonardo da Vinci covertly dissected people as reference for his popular Vitruvian Man drawing (just when you thought that thing couldn't get creepier).

The Catholic Church even got in on the action every so often. In 1308, four nuns took it upon themselves to autopsy the recently deceased (and extremely holy) Sister Chiara of Montefalco, searching for signs of saintliness. According to the not in any way questionable amateur autopsy report, a crucifix was found in her heart, as well as three gallstones in her gallbladder which were thought to represent the Holy Trinity.

A couple centuries later they were still at it—in 1533, officials of the Church ordered the dissection of conjoined twins after their death, in order to determine if they shared a soul (in case you're wondering, the verdict was no, since each girl had her own heart which, at the time, was thought to be where the soul was located).

Gosh, if there's anything cuter than the church trying to do a science I don't know what it is. Listen folks, that wasn't quite science, but hey, don't give up, you'll get 'em next time!

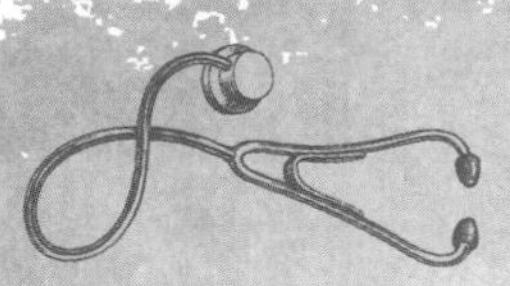

Sydnee's Fun Medical Facts

Though the Catholic church was generally pro-autopsy, Pope Boniface VIII did muddy the waters a bit with one edict. The background is that from the 9th to 13th centuries, it was common for bodies to be dismembered and sometimes boiled until nothing remained but bones for transport and burial. The practice probably began as a convenient and relatively sanitary way to return the remains of soldiers who died far from home. However, by the 1200s there was a veritable fad amongst nobility for having their bodies divided up and buried all over the place. Richard I of England probably had the most complex burial in 1199—his heart, brain, blood, entrails, and the rest of his body were all buried in different places.

Pope Boniface VIII, meanwhile, found this whole practice disgusting, and in 1299 passed a papal bull threatening excommunication for anyone who assisted in this practice, stating that anyone who had requested such dismembering would be denied a church burial (or seven).

This bull may have led to a chilling effect for scientific autopsy at the time, but the edict was not aimed at scientific dissection. That said, it was too weird a story not to share.

BODY OF EVIDENCE

By the time of the Enlightenment, dissection fever was sweeping the Continent. Operating theaters became a popular way for students to observe medical procedures—and were also open to the paying public. (We're retroactively sorry about giving the Middle Ages a hard time.) This was a little ghoulish, but it also indirectly led to the one *Seinfeld* where Kramer drops a Junior Mint into that guy mid-surgery, so . . . that's a wash?

England lagged behind the rest of Europe when it came to dissection, forbidding the practice outright until the 16th century. After autopsies became legal, only ten were permitted per year and only certain members of the Royal College of Physicians and Company of Barber Surgeons could attend. This was obviously a tremendous limitation for anyone wanting to study medicine, and there was plenty of outcry from doctors, students, and natural philosophers (aka scientists).

JUSTIN VS SYDNEE

SYDNEE Protests from the medical community inspired Parliament to pass the Murder Act of 1752, which allowed physicians to dissect the bodies of murderers after they were executed. It was, in fact, an added punishment for certain crimes deemed "punishable by dissection."

JUSTIN And thus ends the most death-metal paragraph it has ever been our privilege to write.

EVERYBODY WANTS SOME BODIES

The Murder Act no doubt helped a bit, but like an unstoppable zombie horde, physicians still craved more bodies. Some were even desperate enough to get a little . . . creative. Creative like grabbing a shovel and heading for the graveyard—or more often dealing with middlemen who did just that. The peak season for grave robbing was November to March when it was cold and the bodies were better preserved. Corpse traffic at one time was so brisk that medical schools had secret entrances constructed for receiving the bodies discretely.

Over time, people got wise to these practices and began to take precautions. Families would hire grave guards, stand watch themselves, or build a cage around the grave called a "mortsafe." Sometimes a family would keep the body at home until it was too decomposed to be worth stealing.

The poor were at greater risk, but everybody was fair game. No, really, everybody: At the Ohio Medical College, the stolen body of U.S. Senator John Scott Harrison (son of President William Henry Harrison) was discovered by visiting dignitaries . . . his son and nephew.

DIY CADAVERS

The corpse business was in so profitable, it should come as absolutely no surprise that opportunists eventually wised up, and started creating the cadavers themselves. The most famous such criminals were William Burke and William Hare of Edinburgh, who killed sixteen people and sold their bodies to a Dr. Robert Knox before they were caught and turned into a feature film

Febreeze, if you're out there and ever want to get hardcore with your marketing, I just came up with an amazing pitch. I know, I know, it may seem extreme, but we're gonna keep it real tasteful. I've even got a slogan all picked out! "Febreeze: Save the Memories, Ditch the Stink." . . . Fine, I'll workshop it.

starring Simon Pegg and Andy Serkis. Oh, and also hanged. Burke was not just hanged, but dissected and displayed as a crime deterrent after. In addition, his skin was made into a purse which is still on display at the Police Museum in Edinburgh. But hey, a movie!

In an attempt to stop this creepiest of crime waves, Great Britain passed the Anatomy Act of 1832, which allowed all executed and dead criminals (even nonmurderers) to be used as anatomical dissection specimens. Similar laws had already been passed in the United States during the late 1700s, allowing for dissection to be added as a punishment for murder.

With all this, grave robbing persisted in the United States as the growing country desperately needed doctors, who had to dissect the human body to finish training. Tension between doctors and wary patients and their families built until 1788 and the Doctors' Riot.

I PREDICT A RIOT

In the late 1700s, Columbia College housed New York City's only medical school. It was also near the Paupers' Cemetery, where the poorest New Yorkers were interred, sometimes several to a grave, as well as a cemetery reserved for slaves. In the winter of 1787–88, the number of grave robberies reported in the press rose sharply. There had been calls, especially among the poor and freed slaves, for corpse-theft laws to be more strictly enforced. It was in this contentious environment that medical student John Hicks, Jr. found himself dissecting a body in his lab at New York Hospital. When he noticed some kids playing outside trying to peek into the lab, Hicks picked up the cadaver's arm and waved it at the kids to scare them. He took the joke a little too far when he added that it was their mother's arm he was holding and he'd slap them with it if they didn't scram.

The kids were obviously terrified, but the anecdote might have been lost to time if not for a few unfortunate coincidences. One of the children was particularly upset because his mother had, in fact, died recently. That prompted the boy and his dad to travel to Trinity Church graveyard where, by chance, they discovered her grave had indeed been robbed. (It's perhaps also worth noting Hicks was studying under Dr. Richard Bayley, a physician originally from England who had been known to steal a body or two in his day and who regularly had his students, like young John, do the same. So it wasn't wild for assumptions about his illicit activities to be reached.)

The boy's father, now irate, went around the neighborhood stirring up a mob in response to this perceived atrocity, and they all marched down to the hospital to confront the doctor.

JUSTIN VS SYDNEE

JUSTIN And now you can buy a corpse on any street corner with no waiting period or background check. They're often offered as complimentary gifts when you buy really fancy luggage. Heck, we're drowning in the things over here, last summer we donated a bunch of them to a haunted house . . . Sure do miss those corpses.

SYDNEE Well, no, there's still a—

JUSTIN You know what, Syd, I'm gonna go buy those corpses back. We'll make room! What a time to be alive! And dead!

Despite its noble intentions, the mob quickly spiraled out of control. Around a hundred men stormed the hospital and destroyed the anatomy lab. They also raided the home of a Dr. Sir John Temple, for no other obvious reason than that his name sounded like surgeon. As the crowd gained mass and momentum, they beat up as many medical students as they could find and, as a result, many doctors and students had to be jailed for their own safety. One medical resident hid in a chimney.

The next day, the mob had increased to 5,000 and they stormed Columbia College as well. They took off for the jail, wanting the students being kept there turned over to them. At this point, the Governor dispatched the militia. The mob was unmollified (again, a mob) until one of their number hit a baron who was trying to stop the riot in the head with a lobbed rock. At that point the shooting began; when the smoke cleared, as many as twenty people had been killed and several more wounded.

Luckily, there were a lot of doctors on hand.

THE AFTERMATH

This actually triggered riots in other major cities, seventeen in total, over the practice of grave robbing. Eventually the riots had the desired effect and some grave robbers were forced to stand trial (and face stiffer penalties). This crackdown also led to the passage of "bone bills," improving methods for a student to legally obtain a corpse.

DO WE STILL DO THIS?

Well, if we're talking about stealing bodies for dissection, thankfully it's no longer necessary thanks to the generous individuals who donate their bodies for research. As a first-year medical student, there is no more humbling experience than your first day in anatomy lab. The tremendous gift given by these kind people is the cornerstone of my, as well as every other practicing physician's, medical education.

Corpse theft, however, is still alive and well (pardon the pun) thanks to skeletons. Bodies donated for dissection frequently have the skeleton destroyed in the process, so there's a brisk black market for intact skeletons. India has long been the world's leading exporter of grade-A skeletons, despite the export of remains being outlawed in 2007.

STEAL A CORPSE THE OLD-FASHIONED WAY

An 18th-century physician hankering to dissect a body had a few extralegal options. You could grab a shovel and go the DIY route. You could take your chances with some no-name grave robber. Or you could hire the very best — which, in this case, meant the Resurrection Men. These snappily dressed corpse-thieving pros could snag a body in under an hour — so fast it barely had a chance to stop breathing. If you're thinking that doesn't sound so tough, you've clearly never taken the Resurrection Man Challenge. Which is good, because it's illegal and gross. But if you were going shovel-to-shovel with the grave-robbing experts, here's how it would go down.

1

You'll need: A fresh grave, some attractive mourners to distract any passersby, a tarp, shovels, a hand drill, and a chain with a hook at one end.

Choose a nice fresh grave. The dirt will be easier to dig up and restore, but more importantly, fresh is always your go-to when exhuming bodies. All set? The clock starts when you spread a tarp near the head of the grave.

2

Carefully excavate just the top of the gravesite, shoveling the dirt onto your tarp. You just need to expose the first foot or two of the coffin lid. Now grab that drill and make a bunch of little holes in the exposed part of the lid to weaken it.

3
Whack the weakened portion to crack it open and use your chain-and-hook setup to grab the body by its shroud. (Shrouds were all the rage back then, conveniently enough).
4
Pull the body clear, lay it on the grass, and stuff the shroud back into the coffin. Now dump all the excavated dirt off the tarp back into the grave and pat it all back into shape.
5
Move the body onto the tarp, and check your watch. If this took you an hour or less, you're ready for the big leagues.
PRO TIP If you need to take public transportation to your next destination, ditch the tarp and pass the stiff off as a drunk friend. That's how the real-life Resurrection Men did it—like *Ocean's 11* with a *Weekend at Bernie's* twist at the end.

Opium is derived from the poppy plant (that scene in *The Wizard of Oz* makes so much more sense now, right?)—specifically the *Papaver somniferum* species.

As far back as 3400 BCE, ancient Sumerians were smoking and eating poppies, which they called *hul gil*, or the "Plant of Joy." They weren't even pretending to use it medicinally though; it was all for funsies. For the next couple thousand years, opium spread to the Assyrians then to the Babylonians and Egyptians. Again though, their opium use was all for laughs. By 460 BCE, Hippocrates (the "First, do no harm" guy) was suggesting it for pain, "women diseases," bleeding, and a boatload of other uses. It's a wily ingredient that changes form throughout the years, like laudanum, a blend of opium and alcohol that emerged in the 1500s. But by 1803, German scientists isolated the active ingredient of opium, calling it "Principum somniferum." These days we just know it as morphine.

By 1874, morphine was used to create heroin, and if you're curious how that worked out, we're assuming you haven't watched the news in a while.

USED TO TREAT:

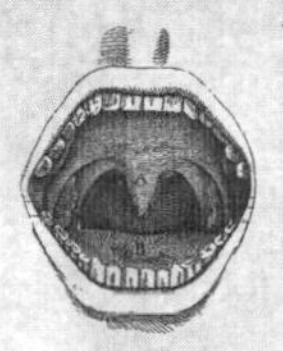

TEETHING

Dr. Farney's Teething Syrup was a delicious, completely kid-safe (wink) blend of alcohol, morphine, and chloroform. Did it work? Well, the kid wasn't complaining!

GASSY BABIES

Baby's got a case of the toots? Well, you're going to need a whole other kind of opium, specifically Dalby's Carminative, a late 1700s bit of snake oil for "infants afflicted with wind, watery gripes, fluxes, and other disorders of the stomach and bowels."

"FEMALE TROUBLE"

Morphine and laudanum were frequently used to treat "hysteria" (old-timey code for "my wife keeps disobeying") but also real problems like gynecological issues and menstrual pain. Part of the reason for their popularity in 1800s America was the largely women-led temperance movement, which warned against the dangers of alcohol but was just fine with opium. As the 1800s ended, two out of three morphine addicts in America were women.

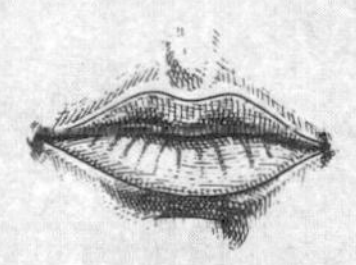

BEAUTY CONCERNS

In the 1800s, it was very chic for women to look frail and sickly, and what better way to achieve that look but a nasty opium addiction? Unless you wanted to get tuberculosis.

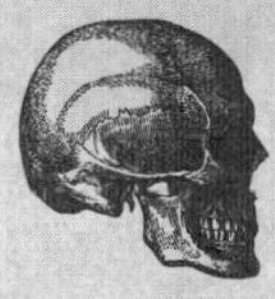

SEVERE DEHYDRATION

Godfrey's Cordial (opium, spices, and treacle) was used by parents and nurses to treat a wide variety of symptoms, but this was perhaps the most upsetting one.

INSOMNIA

Patent medicines like Godfrey's Cordial and Atkinson's Royal Infants Preservative were sold by the gallon to parents and nannies. Though technically sold for medicinal use, it was used by so many parents to keep kids sleepy and docile it was euphemistically known as "Quietness." The tragically predictable outcome: Infants frequently died from overdose or lack of appetite. Okay, so severe dehydration was the second most upsetting.

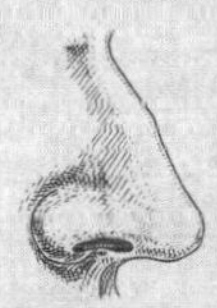

COUGHS AND COLDS

Dr. John Collis Browne had a secret ingredient for his cough and cold remedy Chlorodyne: cannabis tincture. He even added a little chloroform to the potion, which was originally designed to treat cholera. Chlorodyne eventually ditched the cannabis, but weirdly that wasn't enough to ward off the addicts it left in its wake.

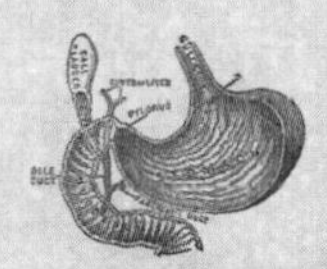

DIARRHEA

Fun fact! Opiates constipate you, so this frequent opium-based patent medicine claim is probably on the level! (Note: There are many better ways of treating diarrhea. Don't be a goofus.)

An Electrifying Experience

One day in the late 1700s, an Italian physician, physicist, biologist, and philosopher named Luigi Galvani was in one of those weird moods where you're just determined to do some sort of science. Anything will do when you're in a mood like that. Galvani had been stroking a piece of frog skin to generate static electricity, as ya do. When he picked up a scalpel to keep dissecting that frog, it delivered a shock to the dead frog's leg. The leg kicked, and Galvani realized at once that he had unlocked the secret to reanimating the dead.

The Innovator

Well, okay, Galvani didn't think that exactly. In fact, the reportedly very modest and pious Galvani had his head screwed on pretty straight for a late 1700s scientist. An anatomy teacher and researcher with degrees in medicine and philosophy, Galvani was fascinated by the recently-discovered connection between life and electricity. Hence the frog experiments, which led to his discovery that muscles contract when stimulated by electricity, which we now call galvanism. Galvani thought this was caused by he called "animal electricity." Before Galvani, scientists had believed that muscles operated via movements of fluid and air.

The Treatment

A couple of decades later, Galvani's nephew Giovanni Aldini decided to take up the family business with a cool new spin—reanimating human corpses. He gave a number of public demonstrations on heads and corpses, most notably in 1803 on the corpse of murderer George Forster, which Aldini purchased shortly Forster's hanging. Through the careful application of electricity, Aldini was able to make the body contort its facial muscles, and even raise a hand eight inches off the table. According to one account, the display was so ghoulish that a surgeon named Mr. Pass who witnessed it died of fright shortly after returning home.

How'd That Work Out?

Perhaps the best way of answering that is to reveal that this entire section has been written by me, George Forster. I'm 240 years old now and—no, no, it's still Justin and Sydnee. George Forster and every other corpse subjected to galvanic reanimation is still deader than disco. By the 1940s, we'd learned we could fix an irregular heart rhythm using the electrical charge from defibrillator paddles—but contrary to popular belief, they can't restart a totally stopped heart.

FUN FACT: Some thirteen years after Aldini's exhibitions, a writer named Mary Shelley and her friends, the poets Lord Byron and Percy Shelley, spent an evening discussing the possibility that galvanism could be the key to reanimating the dead, and what the moral implications would be. As you might have guessed, this inspired her to write her classic novel, *Frankenstein*. That said, Shelley's Dr. Frankenstein doesn't actually use electricity that much in the book. The unforgettable image of shooting sparks reanimating Boris Karloff in the 1931 film was the brainchild of set designer Kenneth Strickfaden, who got his start as a carnival electrician.

FAT IS FOLLY

when it can be reduced easily, conveniently and best of all, **Safely**, by the use of

La Parle OBESITY SOAP

This **Obesity Soap** (used like an ordinary soap) positively reduces fat without dieting or gymnastics. Absolutely harmless, never fails to reduce flesh when directions are followed.

Send for book of testimonials. Box of 2 cakes sent prepaid on receipt of **$2.00.**

Norwood Chemical Co., St. James Bldg., N.Y.

Weight Loss

It's amazing what people have done over the years to avoid just exercising and laying off the cheese puffs.

The history of humanity, or at least a few hundred years of it, is the inspiring story of our people trying to find ways of dropping pounds while avoiding better eating habits and exercise. "There just has to be an easy way that works for everyone!" we've told ourselves for generations, and even though we haven't found it yet, we have never and likely will never stop searching.

In the early days of human history, staying fit was pretty easy. When you've got to hunt and gather all of your sustenance, the idea of working to lose weight was largely unheard of. However, as time went on and people had regular access to plentiful food sources, the necessity to balance intake and expenditure of calories arose. What follows are some of the most notable, outrageous, tragically ill-advised, and just plain weird ways we've tried to slim down without, you know, trying too hard.

DRAIN THE SWAMP

The theories behind weight gain have been as strange as the dieting methods themselves. One proposed by Dr. Thomas Short in his 1727 book, *The Causes and Effects of Corpulence,* was that obese people often lived near swamps. An odd leap in logic? You bet. But also deeply offensive to marshy guardian Swamp Thing, who we can all agree is pretty swole for a guy who lives off chlorophyll.

On the bright side, Short's hypothesis lead to perhaps history's simplest diet plan: Don't live near a swamp.

DEPRIVATION CHIC

In the 1800s, people got less creative but more direct in their battle of the bulge. The prevailing wisdom was simply not to eat anything if you wanted to lose weight. Lord Byron became a diet icon of the period, weighing himself often and working hard to maintain a slim figure. At times this meant only consuming vinegar and potatoes, or perhaps soda water and biscuits. He also wore heavy clothing to promote excess sweating and would take laxatives frequently. It really throws his lovely work into a new light when you imagine how frequently the crafting of it must have been interrupted by bathroom breaks.

This fad became popular among men and women alike, depriving themselves of most nutrients and trying to subsist on as little as possible while maintaining the waifish look that had become trendy at the time.

LOW-CARB'S DEBUT

The first mention of an Atkins-like high-protein and carbohydrate-control diet actually dates back to 1863 in a *Letter on Corpulence: Addressed to the Public* by former undertaker William Banting. Dieting had become pretty popular at this point, but many people were relying on simply decreasing the amount of food they ate in their attempts to lose weight. Banting, obese yet unwilling to starve himself, cooked up his own plan. He advised eating mainly meat with the occasional glass of sherry, and completely eliminating potatoes, bread, milk, sugar, champagne, and beer. It worked for him—he managed to lose fifty pounds with this strategy, and he built a significant following. In fact, his pamphlet is still available online, and the term "banting" was used to mean "dieting" for over a century.

CHEW YOUR WAY SKINNY

When Horace Fletcher, a 19th-century American art dealer, was denied life insurance due to his weight he was struck with the idea for one of the strangest diet fads, later known as Fletcherism. The big secret? Just chew! No, no, chew a lot . . . okay, now keep chewing . . .

Before you wear out your jaw, you should know Fletcher advised chewing food one hundred times before swallowing. Even liquids. Somehow.

The way Fletcher saw it, you'd eat less because you'd run out of time to eat, and burn a bunch of jaw-exercise calories in the process. Fletcher became known as The Great Masticator, and roped in plenty of the time period's luminaries like Upton Sinclair, Henry James, and John D. Rockefeller.

It got weirder. Fletcher claimed his program had benefits beyond just weight loss. He insisted, most notably, that followers of Fletcherism would need a bowel movement just twice a month and furthermore, that this bimonthly deposit would smell like warm biscuits.

Perhaps the most surprising element of Fletcherism? It may not be totally bogus.

One study (with a small sample size that necessitates further study) suggests that he may have been on to something. Well, not the biscuit part. Obviously.

OH RIGHT, CALORIES

Dr. Lulu Hunt Peters introduced the concept of calorie counting in her book *Diet & Health: With Key to the Calories*. Her book was largely aimed at women, and revolved around the idea that dieting not only lead to a thin figure, which was by now very popular, but also better self-control, health, and well-being. In a shocking about face in the history of dieting to this point, she based her calorie recommendations and formulas on actual research (alongside her own anecdotal results).

She tried to inspire women by insisting that this was even a patriotic duty. Her book was published during World War I, a time of rationing. Her argument was that every bite of food of which a woman deprived herself was another bite for her hungry children. To ease the pain of sticking it to Fritz via cookie deprivation, she also advised that women start "Watch Your Weight Anti-Kaiser Classes."

PUFF THE MAGIC DIET

If women found they couldn't stick with Peters' plan, they were offered a far easier-to-follow diet secret by . . . cigarettes. Lucky Strike launched an advertising campaign in 1928 aimed specifically at women who were struggling to achieve whatever was deemed an "ideal figure" at the time.

Referencing the widely held belief that smoking would help one lose weight, they began to run the tagline, "Reach for a Lucky instead of a Sweet." The campaign was, heartbreakingly, very effective—and was thwarted only when the candy industry began to threaten litigation.

GET A PARASITIC PAL

After observing that tapeworms are very good at stealing nutrients from its host, some folks theorized that maybe it would be good at taking

As a bonus fun fact, this catchy slogan was almost certainly inspired by "Reach for a Vegetable instead of a Sweet," a tagline for Lydia E. Pinkham's Vegetable Compound an 1870s herb-and-alcohol-based tonic for menstrual pain, some variation of which is still sold to this day. Although with less alcohol (the Prohibition-era tonic was a bracing 40 proof), and no unicorn root at all.

away those unwanted pounds too. Tapeworms, if ingested, do indeed attach to the wall of your intestine and just sort of hang out, eating what you eat, for maybe your whole life.

Good deal right? Not so fast! First: Eww, gross. Second, and more important: Though they may well cause some appetite suppression or diarrhea, which could lead to weight loss, they are fairly small and not up to the task of absorbing a typical human's daily caloric intake. So don't expect big results.

Besides, while pills containing tapeworms were certainly advertised for the purpose of losing weight, it is not clear that these capsules actually contained any parasites. That sounds shocking, but consider the diet pills you've seen advertised in the last week and it'll seem distinctly less nutty.

We're sort of burying the lede here: This is a very dumb thing to do. While, yes, you can buy tapeworm pills in some parts of the world, it is extremely dangerous. You can't be certain what kind of parasite you may be ingesting and some of these infections can be fatal.

This wasn't advice we expected to offer in our very first published book, but here we are: Don't eat tapeworms.

THE MIRACLE FRUIT

Have you ever wondered why all those people in '80s movies are always eating grapefruits for breakfast? Sure, it could just be because they liked to eat grapefruit, but we refuse to believe that anybody on the planet really likes grapefruit, so that strikes us as profoundly unlikely.

Much more probable is that they were inspired by a fad diet that actually dates back to the 1930s. It originated among the stars of Hollywood and was based on a belief that grapefruits contain special enzymes that would help burn fat faster. When the diet was unfortunately resuscitated in the '70s, it was often called The Mayo Clinic Diet, despite no connection whatsoever with Mayo Clinic. In hindsight, that should have been a pretty good tip-off of the plan's bogusosity.

Variations of the diet ranged from eating half a grapefruit with every meal, to eating essentially only grapefruit with an occasional side of some sort of meat. Proponents claimed you could lose ten pounds in ten days. What an amazing result! (Psst: Did we mention that you also need to limit yourself to 800 calories a day?)

CABBAGE TAKE THE WHEEL

While the grapefruit diet lay dormant for a few decades, a new and possibly even less appealing plan took center stage in the 1950s: The cabbage soup diet. Instead of a grapefruit, you eat at least two bowls of cabbage soup a day along with various fruits, vegetables, and meats for a week.

Each day involves a slight variation in what you can take in with the soup, just to keep things interesting. For example, on day four of the diet, you can eat six to eight bananas and all the skim milk you want in addition to your soup. Presumably, that would be followed up with pooping forever and ever.

The diet only lasts seven days and obviously has some drawbacks, including being very hungry and smelling like a fart collection that had been stored for a month in a gym teacher's trunk. You will almost certainly slim down though . . . since you're restricted to 800 to 1,000 calories a day.

ME GENERATION WANT COOKIE

Dr. Sanford Siegal introduced The Cookie Diet in 1975. We know, we were tempted to enroll

based on the name alone, but maybe hold off for a paragraph or two.

Siegal was already working as a weight loss physician when he came up with a blend of amino acids that he thought would help keep his patients feeling full. He put them into a cookie that patients could eat six of (good for a total of around 500 calories) throughout the day followed by a 300 calorie "sensible dinner" (read: skinless chicken and vegetables). Though Siegal is no longer in the game, he spawned many copycats and you can now find health food stores littered with versions of the cookie diet, along with celebrities and athletes willing to endorse them.

If you've done a little math in the last paragraph, you've probably figured out that cookie dieters are limited to 800 calories a day, a dangerous amount no matter how many cookies you get to eat.

MYSTERY MEAT

The Last Chance Diet was unveiled in 1976 by Dr. Roger Linn in a book by the same name. It was arguably the most dangerous plan we've covered yet (and that's saying something). At its core was a proprietary drink of Linn's own creation called Prolinn (get it?). Though he referred to it only as a "protein blend," investigation of the liquid revealed that it was made out of hooves, bones, horns, hides, tendons, and other slaughterhouse byproducts.

Okay, that's very gross and bad, but hey, you've probably eaten a hot dog, so let's maybe simmer down a bit. No, the real problem was that dieters were restricted to 400 calories a day and Prolinn was essentially devoid of major nutrients and vitamins. The Last Chance Diet proved to have an eerily prophetic name. After around 60 of its adherents died suddenly while attempting the diet, it fell out of favor.

EVERY BREATH YOU TAKE

No one takes calorie restriction to the extreme, though, like the Breatharians. A fad that originated in the 1980s, Breatharians believed that once you attain harmony with the universe, you only need air—no food, no water, just air.

Many have claimed that they have gone for extended periods of time without any actual sustenance, some up to seventy years. However, proving that on camera or in person has been . . . challenging. One advocate attempted to demonstrate it on *60 Minutes*, but almost died of dehydration after four days.

Tragically, at least five people have died trying to follow Bretharian teachings, which are so obviously nonsensical we hope you don't need us to walk you through it. The short version: Food good. Eat to live.

WHAT HAVE WE LEARNED?

Don't get too smug, present-day readers; a lot of the strangest diets were still to come. Consider, for example Dr. Peter D'Adamo's 1997 book *The Blood Type Diet,* which features eating plans for each blood type and . . . not a whole lot of science. The 2000s brought *The Vision Diet,* which advised dieters to don blue glasses so that food looked less appealing. You can even buy a seaweed soap on Amazon that makes you lose weight! (It does not.)

Here's the skinny: Fad diets may work for a little while, but they're almost never sustainable, and occasionally also pretty dangerous. If you want a surefire way to lose weight, make better food choices, get more exercise, and smoke as many cigarettes as you can get your hands on.

We're kidding, obviously. A couple packs a day should be plenty.

Okay, Sydnee looks pretty mad at me at this point, so I'm going to stop the chapter here.

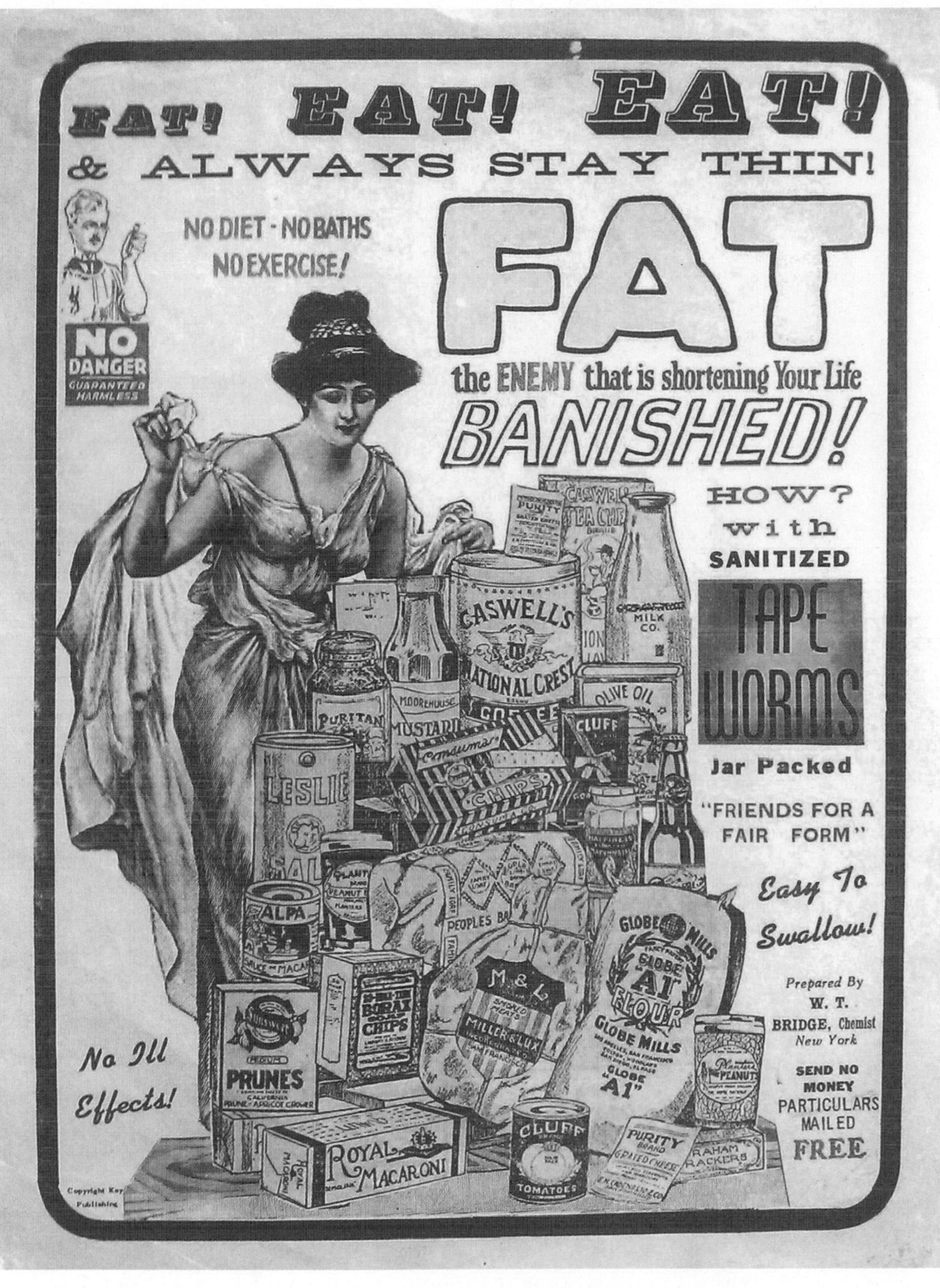
EAT! EAT! EAT!
& ALWAYS STAY THIN!
NO DIET - NO BATHS
NO EXERCISE!
NO DANGER
GUARANTEED HARMLESS
FAT
the ENEMY that is shortening Your Life
BANISHED!
HOW?
with
SANITIZED
TAPE WORMS
Jar Packed
"FRIENDS FOR A FAIR FORM"
Easy To Swallow!
Prepared By
W. T. BRIDGE, Chemist
New York
SEND NO MONEY
PARTICULARS MAILED FREE
No Ill Effects!
CASWELL'S NATIONAL CREST COFFEE
PURITAN
MOOREHOUSE MUSTARD
OLIVE OIL
CLUFF
LESLIE SALT
CHIPS
ALPA
PEOPLES BA
GLOBE MILLS
GLOBE "A1" FLOUR
M & L
MILLER & LUX
PRUNES
BORAX CHIPS
ROYAL MACARONI
CLUFF TOMATOES
PURITY GRATED CHEESE
GRAHAM CRACKERS
MILK CO.
Copyright Key Publishing

MIRACULOUS UNIVERSAL CURE-ALL

CHARCOAL

From the barbecue
to the medicine cabinet

Summer's almost here . . . or is here
. . . or just ended . . . or seems like forever
away. Listen, we have no idea; we're stuck in this
book. Summer is a time, okay? The important thing is
that somewhere in the world, at some point, it will be hot
outside. When that time comes, there are going to be a lot
of grill masters, their bellies all full of fresh-charred meat,
who all share the same question: "That was a very good steak,
but now what am I gonna do with all this medicine?"

No dear reader, we didn't let a typo slip through the cracks:
That big old pile of briquettes left over after the searing is done
is just a big, still surprisingly warm pile of healing.

The use of charcoal as medicine—specifically, activated
charcoal—dates back to ancient times. This is a form of charcoal
that has been chemically or physically processed in order to create
more surface area and small pores that can more readily adsorb
other substances. The tricky bit is that charcoal really is good for
some things, but the problem in medical history is that
substances that actually help with some issues are
for some reason expected to help with all issues.
We bet that kind of pressure is exhausting.

USED TO TREAT:

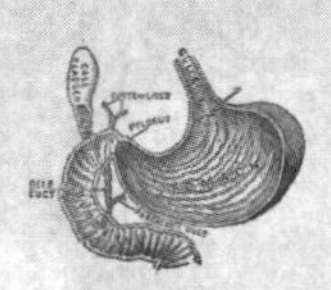

DIGESTION

For literally thousands of years, we've been eating charcoal to help with tummy trouble. Ancient Egyptians did it for gastrointestinal distress, Hippocrates swore by it as a stomach aid (as well as a treatment for vertigo, anemia, and epilepsy). Even doctors in the 1800s used it for tummy troubles such as diarrhea and flatulence.

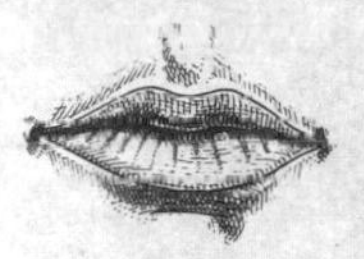

COLD SORES

This was a Pliny the Elder original, another cure that required you consume the charcoal. (It's a really good thing that Pringles were eventually invented, the ancients were desperate for snack foods, apparently.) Oh, it also supposedly would cure your carbuncles.

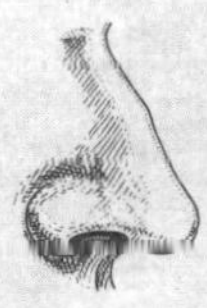

BAD ODORS

When Egyptians had a little charcoal left over from treating their digestion, they used as a stink repellent, especially for open infected wounds.

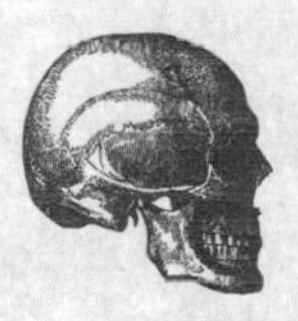

WATER PURIFICATION

By 400 BCE, Phoenician sailors were carrying drinking water in charred barrels to preserve it, a practice that was used on long sea voyages until the 1800s. This is a good one! Active charcoal carbon filters can remove chlorine, sediment, and weird flavors and smells from water. If you've ever used a filtered drinking bottle, you very well may have been slurping your water through charcoal.

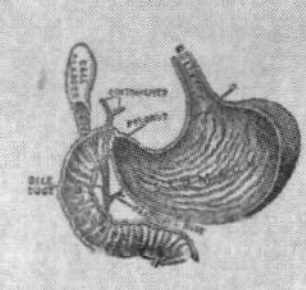

BLEEDING

Charcoal had sort of a resurgence in 1800s Europe. We've covered the GI treatments, but it was also really popular as a way to stop bleeding. It was applied topically, which makes a certain sort of sense, we guess, but also orally. That's right, they believed if you ate some charcoal your nosebleed would stop.

In the late 1800s, charcoal was prescribed for rectal bleeding, as per this journal entry from John Fyfe M.D. "The specific use of charcoal is to arrest hemorrage from the bowels. It has been used in enema, finely powdered to four ounces of water, thrown up the rectum. Why this checks it I cannot tell; that it does it, I have the evidence of my own eyes."

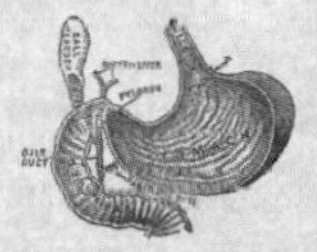

POISONING

In the 1900s, Japanese and French surgeons started using activated charcoal to treat poisoning. This, as it turns out, was very much on the level. If you remember the water purification bit, charcoal is very good at soaking up weird stuff. The National Capital Poison Control Center supports the use of activated charcoal in many poisoning situations. The Center also says that one teaspoon of activated charcoal has the same total surface area as a football field. Which . . . well, we just thought that was neat.

FUN FACT: Justin asked Sydnee what "carbuncles" were while writing the above paragraph and she said it was sort of a "multi-headed boil" and that was probably enough info but then she showed him a Google image search result for "carbuncles," and so he died and now will have to write the rest of the book from Hell.

The Black Plague

If we asked you the scariest thing a disease can do, you'd probably say "kill you" and in a sense of course, you'd be right.

But after years of studying the history of disease, we're a little less impressed by the run-of-the-mill fatal stuff. After all, anything can kill you! Just look around. See that loose screw in the ceiling fan? That blender just a shade too close to the bathtub? That tuna salad you left out in the sun? Death is waiting for you around every corner.

What really gets us excited these days is those diseases that don't just kill people . . . they change the very course of history. And the plague is just such a disease.

Just how scary was the plague? It's so scary that "the plague" isn't a scary enough name; we started to call it "The Black Death," right around the time it wiped out half of Europe. Since the very first plague bacteria infected its first human host, this disease has wreaked absolute havoc on our species. Or at least it did right up until the discovery of antibiotics blunted a bit of the threat. Nonetheless, it has earned itself a place in the hall of fame of history-altering illnesses, which isn't a real place but probably should be.

The plague is caused by a bacteria called *Yersinia pestis*. Plague comes in more than one flavor, with multiple forms each of which affects different systems and organs in our bodies and has its own complications and mortality rates. If you've heard of just one type of plague, it's probably the bubonic one, which gets its name from the impressively swollen lymph nodes known as buboes. These could appear in a sufferer's groin or armpit and were often large, tender, and hot. While these days we use "the plague" and "bubonic plague" interchangeably, other forms can actually be more deadly. Fun!

While we're pondering the relative scariness of nomenclature, could we not cook up something a little more threatening than "buboes"? Big, hot, gross groin bubbles and we give them a name that would be more appropriate for an adorable Super Mario enemy? You can do better, science.

A PLAGUE OF PLAGUES

If you're not into buboes, or just want to be different, you might consider septicemic plague for your next deadly disease. This variant spreads through your bloodstream and can cause bleeding from your mouth or rectum, shock, and even gangrene of your fingers or toes.

Still not convinced? How about the pneumonic plague, which helps you remember large chunks of information with fun acronyms? No, no, actually it primarily affects lungs, and is often mistaken for a really severe case of pneumonia. While its symptoms may sound the least dramatic, this one is actually the most fatal form, and needs to be treated as soon as possible.

Come to think of it, there's not really a "good" plague to catch.

The plague is not an absolute death sentence, but . . . it ain't great. Without treatment, about 50 to 90 percent of infected people will die. With modern treatment, that number is reduced to a much-better-but-still-not-all-that-comforting 15 percent.

A BRIEF HISTORY OF PLAGUE

We know now all these details thanks to, well, science. Folks in history weren't so lucky—which we know because the plague is mentioned as far back as the Bible. In the first book of Samuel, the Philistines are struck with some sort of tumor-causing plague after stealing the Ark of the Covenant from the Israelites. (For more details, seek out the popular documentary *Raiders of the Lost Ark*.) Descriptions of this disease, as well as the outbreak suffered by the Israelites after the Ark's return, all seem to point to plague.

In 532 CE, the Great Plague of Justinian spread from Byzantium throughout the Middle East, around the Mediterranean, and over to Egypt, eventually killing some 25 million people. This death toll represented 50 to 60 percent of the Eastern Roman Empire—and at least 13% of the entire world's population. For the next few centuries history records only minor outbreaks, but that luck didn't hold. In 1328, the plague appeared in China, quickly spreading along trade routes to Italy, and then to the rest of Europe. This outbreak lasted until about 1351, and wiped out 30 to 60 percent of Europe's population.

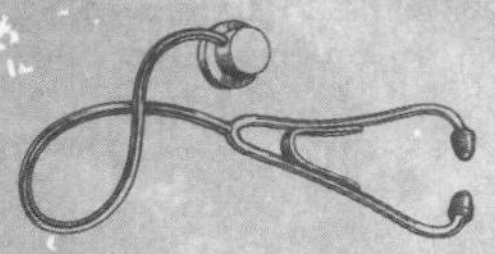

Sydnee's Fun Medical Facts

Technically speaking, the Black Death is only used to refer to the outbreak of plague that devastated much of the world, around the middle of the 1300s. That's what historians will tell you. That said, the name "Black Death" is just too sexy for the media to pass up, so it's frequently used as an alternate name for the plague wherever it occurs.

You hear stats like that all the time when you read about plagues, but take a moment to let it sink in. Half of everybody you've ever known, gone. Entire towns wiped out, sometimes so quickly that the bodies were left stacked in town squares. We try to keep things light here, but yikes. It's unfathomable. It's horrifying. It's the Black Death.

SOMEBODY CALL FOR A DOCTOR?

Around the 17th century, we started seeing references to a new kind of specialist, to use the term loosely. Afflicted cities and villages hired plague doctors, supposed specialists who charged to help treat the sick and prevent the disease from spreading. These physicians were usually second rate, which is saying something, considering what passed for first rate at this point in history. The average plague doctor may or may not have actually been trained in any sort of medicine, but they all shared one important trait that qualified them for the job: they were willing to do it. They didn't have much know-how, and they seriously lacked efficient medicines, but they did have some pretty sweet gear.

IN THE AIR TONIGHT

It's not likely that even the most conscientious plague doctors got much information from their examinations. And even if they had, any theories or findings would have been based on sketchy science at best. We humans have come up with many terrible theories as to why sickness occurs over the years, including magic, evil spirits, vengeful gods, unbalanced bodily fluids, and even the weather. Plague doctors were happy to give you advice (from across the room anyway), but it wouldn't have been very helpful.

The most popular disease theory at the height of the Black Death was that illnesses are caused by miasma, which is Ancient Greek for pollution. The miasma theory of disease is easy to understand: Sometimes air is just bad and gross, and if you breathe the bad, gross air you get sick. Sure, it kinda sounds like an explanation of disease that your not-so-bright four-year-old nephew would cook up while half-distracted by the TV, but hey, they were doing their best.

Miasmas could arise from anywhere, it was believed, but they came to be associated with poor sanitation. This understanding of disease was totally wrong in the most helpful way possible, because it did indirectly lead to improved sanitary conditions that may well have helped stem the tide of the outbreak. For the first time, people cared about removing the human and animal waste that often littered the streets. In addition, the bodies of the deceased were removed from the home, and buried in pits or burnt. Sure, we were on the wrong track—we wouldn't figure out the whole germ thing until the 19th century—but at least the train was getting cleaner.

TAKE TWO LIVE CHICKENS AND CALL ME IN THE MORNING

As we've established, the plague was terrifying, spreading fast and far, and killing the majority of those who contracted it in many places. Amping up the terror was the fact that nobody understood the disease well enough to combat it effectively. But as with all illnesses and, indeed, pretty much every night-impossible challenge in human

The Plague Doctor

Your mom probably wants you to be a doctor when you grow up, right? Booooring. Nobody looks good in a white coat, clipboards are lame, and crippling student debt isn't as fun as it sounds. But wait! What if there was a way to skip all those boring science classes and spend every day dressed like an NPC in *Assassin's Creed*? That's right kids, you too can be a plague doctor. All you need is a spiffy outfit, a bunch of stinky herbs, a nice long stick, and a really strong immune system.

Accessory #1 is a beak-like mask stuffed full of potpourri. The idea was that filtering the bad air through nice-smelling things would help keep you from becoming sick yourself. (Always a great idea!)

Don't forget your examinin' stick! Since no one actually wanted to touch an infected plague victim, poking them with a stick seemed like a decent alternative. You know, just a good old, scientific poke from a dude in *Hellraiser*-worthy leathers and a mask that smells like grandma's linen closet.

Leather gear is treated with wax to make it waterproof. In case you get caught in the rain or splattered with disgusting fluids. Mainly the latter.

The Plague Doctor's hat, robes, leggings, gloves, and boots were made of matching Moroccan leather. Maybe their undies too. We don't judge.

history, our ignorance didn't stop us from trying.

Some of the advice is what you'd call wrong, but relatively harmless. A plague sufferer might have been instructed to avoid meat and cheese, eating a diet of only bread, fruit, and vegetables. Their families would have been told to keep patients in bed, wash them with vinegar and water, and place nice-smelling flowers in the room. All pretty benign stuff.

But when these comparatively chill methods proved ineffective, it provoked the plague doctors to come up with new, worse, more deadly strategies. One of the most popular was . . . lancing the patient's buboes. Remember, we're talking about those nasty, pus-filled swellings that formed under the patients' arms or in their groins. Once the buboes were cut open (probably not with the cleanest of surgical instruments) plague doctors would apply a special medicinal mixture to the wound "to facilitate healing." What would one put on a gaping, still oozing pustule to facilitate healing? Tree resin, the root of white lilies, and human excrement.

Naturally.

Oh, hachi machi does this next part ever get grody. I wish I hadn't read it, and I helped write it. Ugh. Also, if you notice any really obvious typos, it's because I was only half peeking from between my fingers while I was blasting some really soothing Sarah McLachlan.

If that sounds too fancy for your tastes, what with the lilies and all, you may prefer the tried-and-never-true method of simply bleeding the patient. Just cut them and let them bleed a bit and then put some clay and violets over the cuts you made. No problem.

Some sufferers turned to extreme religious practices, such as self-flagellation, in hopes of winning forgiveness and healing from above. Others looked to witchcraft, which offered such treatments as drinking one's own urine or strapping a live chicken to the bubo and waiting for the bird to die. Chickens have had it rough throughout history, but strapped to the oozing, pustule-covered arm of a plague sufferer has still got to be one of the top five worst ways for a chicken to die, right?

An entire field of "pestilence medicine" was developed over the Middle Ages, with fake cures and concoctions aplenty. One such recipe called for the sufferer to roast the shells of newly-laid eggs, crush them into powder, mix that with the chopped-up leaves and flowers of marigolds, put it all in a pot of good ale, add treacle, warm the mixture over a fire, and drink it twice a day. Until . . . well, until you died from the plague.

A WHOLE LOT OF BAD MEDICINE

Finally, when nothing else seemed to work, doctors would just keep making stuff up, hoping to somehow stumble onto a treatment that even remotely made sense. Patients were advised to eat . . . unusual substances, such as rotten treacle, or the powder of crushed emeralds. (Any dirt-poor villager worth their dirt had those lying around, right?) They were told to drink arsenic or mercury, avoid sex, stop going outside, squelch all thoughts of the plague or of death (yeah, that must've been a breeze for a plague victim) and last but not least, abandon their homes to go live in the sewers, where the stench of human waste was supposed to provide protection—somehow. (So much for that miasma theory, huh?)

Of course, sewers are often a place where rats happen to live, by the way. Keep that in mind, because that last bit of advice is about to get way more hilarious.

Fleas like to live on rats, and transmit the disease between their human and rodent hosts, allowing it to spread far and wide. It's easy to see how poor sanitation and overcrowding contributed to the disease's rapid spread, with all those infected fleas hopping around. Rats also

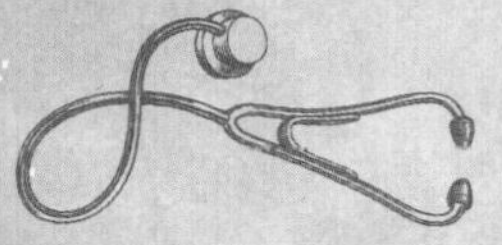

Sydnee's Fun Medical Facts

While no one could figure out how to stop the Black Death, a few enterprising young soldiers saw it as an opportunity for victory.

As the plague swept through Europe and North Africa in the 14th Century, a number of conflicts and wars were also going on because, well, humanity. In this instance, a group of Tatars (descendents of Ghengis Khan's Golden Horde most likely) were laying siege to Caffa, a Genoese port that's now part of Ukraine. The Tatars were already having some difficulty breaching the city's defenses, but things went from bad to worse when the plague broke out in their encampment.

In one of history's earliest examples of bioterrorism, the invading Tatars had the equally brilliant and disgusting idea to toss the bodies of dead plague victims over the city walls. This plan worked, the plague spread throughout Caffa, and the Genoese army eventually retreated.

JUSTIN VS SYDNEE

JUSTIN: OK, perfect, Syd, we've cracked it, it's so simple I don't know why we didn't see it before. We should get rid of all rats. Then we're good, right?

SYDNEE Well, unfortunately, Justin, a lot of animals can get the plague, including cows and sheep and small ground squirrels and rodents that live throughout the western United States, Africa, and Asia. This has sadly lead to isolated cases and small outbreaks of the plague in these areas ever since, many focused in sub-Saharan Africa and Madagascar.

JUSTIN: . . . Right, yes, but I'm still not exactly clear where we ended up vis-a-vis getting rid of all the rats? I feel like we were both feeling really good about it but now—

SYDNEE: We're not getting rid of all rats.

JUSTIN: Fine.

like to live on ships and, since they take their fleas with them, the plague could spread anywhere a ship might sail. Scientists figured this out during the third and last Western plague pandemic in the 1860s, and were finally able to stop the disease from spreading by controlling the rat population.

Rats aren't the only animal hosts, but with the development of antibiotics, we have a variety of medications that are effective in treating the plague. We're able to save 85 percent of those infected, which, though obviously not effective enough, is a heck of a lot better than we were doing in the 1350s.

Thus ends our sobering, chapter-long reminder to hug your local scientist today.

SO, WHAT HAVE WE LEARNED?

Although miasma theory proved incorrect, plague doctors were onto something with their masks and outfits; they were essentially early hazmat suits, precursors to those worn by medical personnel treating outbreaks such as Ebola. (Thanks, plague doctors!) Another important discovery for the modern world is that we now know plague is carried by fleas that live on rats. The ones that live in sewers. (Thanks, plague doctors.)

PLINY THE ELDER

23 TO 79 CE · ROMAN EMPIRE

Gaius Plinius Secundus was born in 23 CE, and died on August 25 of 79 CE. In those 56 or so years, Pliny the Elder was a soldier, navy commander, politician, researcher, naturalist, and probably other stuff he was just too busy to tell anybody about. His magnum opus, *Natural History,* is a massive accounting of all the plants and animals and basically everything in the known world of his time. It was the model for the encyclopedia as we know it. He probably died while trying to rescue a family friend during the eruption of Mount Vesuvius.

We wanted to get Pliny's bafflingly diverse achievements out in front of you, dear reader, in the hopes that they'll buy him a bit of mercy later in this book when we start ruthlessly dunking on him for being wrong about just about everything in the realm of medicine. We try to remember that Pliny was working with a staggeringly primitive knowledge of anatomy and biology. Pliny, notably, did not keep this fact in mind when bloviating about maladies, treatments, and anything else in his field of vision. Don't take our word for it, just check out some of his greatest hits (there were a lot to choose from, but after a while it all starts to sound the same, so we decided on a nice sampling of home remedies you really, really don't want to try).

CATARACTS

For the cure of cataract, the ashes of a weasel are used, as also the brains of a lizard or swallow. Weasels, boiled and pounded, and so applied to the forehead, allay defluxions of the eyes, either used alone, or else with fine flour or with frankincense.

TOOTH CARE

Hollow teeth are plugged with ashes of burnt mouse-dung, or with a lizard's liver, dried. To eat a snake's heart, or to wear it, attached to the body, is considered highly efficacious. There are some among the magicians, who recommend a mouse to be eaten twice a month, as a preventive of tooth-ache.

MENSTRUATION

Contact with [menstrual blood] turns new wine sour, crops touched by it become barren, grafts die, seed in gardens are dried up, the fruit of trees fall off, the edge of steel and the gleam of ivory are dulled, hives of bees die, even bronze and iron are at once seized by rust, and a horrible smell fills the air; to taste it drives dogs mad and infects their bites with an incurable poison.

EPILEPSY

To cure epilepsy, eat the heart of a black jackass, outside, on the second day of the moon. Alternatively, eat lightly poached bear testes, a dried camel brain with honey, or drink fresh gladiator's blood.

WRINKLES

Maidenhair leaves steeped in the urine of a boy not yet adolescent, if they be pounded with saltpeter and applied to the abdomen of women, prevent the formation of wrinkles.

BLOODSHOT EYES

When the eyes are bloodshot from the effects of a blow, or affected with pain or defluxion, it is a very good plan to inject woman's milk into them, more particularly in combination with honey and juice of daffodil, or else powdered frankincense.

Erectile Dysfunction

Hey, we heard your penis wasn't working so great. Luckily, this topic is near and dear to many historic scientists.

Listen, it happens to the best of us. One day, your penis is going for it like crazy, fulfilling all your penis-based dreams and then—kaput. Fear not! The history of medicine has, until recently, been monopolized by people with penises, and this topic has been thoroughly explored.

JUSTIN VS SYDNEE

JUSTIN Listen, I'm going to be straight with you. I'm doing the final edits on this chapter before we send it to our editor, and I've kinda had a rough day, so I'm pretty much going to be writing the word penis as often as is grammatically sound, because I need to cheer myself up and I am apparently eight years old. I'm assuming some of the instances of "penis" will be Lorena Bobbitted out before the book goes to print, but I want you to know I tried. I tried.

SYDNEE I at least appreciate that you're using appropriate medical terminology and not one of the horrifying euphemisms you adopted from your side of the family.

JUSTIN Thank you, Syd. And sorry, reader, that you don't get to enjoy me repeatedly typing "wormy."

SYDNEE Oh, there it is, it happened. We made it forty-three pages, and something in this book has finally grossed me out.

WIENER WISDOM OF THE ANCIENTS

Humans have been trying to come up with treatments for erectile dysfunction for as long as humans have been having erections. Which, as evidenced by the fact that humanity didn't stop with Adam and Eve (or Gronk and Thurkinda if you prefer a less biblical approach), is a very long time.

You've almost certainly deduced this, but "erectile dysfunction" refers to the inability to achieve or maintain an erection. By now, we have learned that it can be caused by a variety of conditions, including high blood pressure, low testosterone, psychological issues, or complications from medications.

Our ancestors didn't know any of that, but that didn't stop them from coming up with ideas for why it happened, and a wide and sometimes terrifying range of treatments. Despite many of these attempts having absolutely no basis in science, and therefore no shot whatsoever of working, humanity has been consistently giving it the old college try since ancient times.

One of the earliest recorded treatments comes from Ancient Egypt, and involved grinding up the hearts of baby crocodiles to rub on the

affected area. Yes, you read that correctly: Baby crocodile hearts . . . ground up . . . to fix our broken wieners. So, the next time you hear of a sweet elderly Floridian being eaten by an alligator, please try to remember that humanity definitely deserves whatever reptilian violence we get.

The ancient Greeks had a simple theory: If it looks like a penis, or is a penis, it might be good for your penis. So, with that in mind, they advised eating the genitalia of roosters or goats to get your mojo working.

Not into eating actual penises? How about a snake? They're kind of phallic and also, bonus, rejuvenate themselves by shedding their skin, so Greek physicians figured hey, maybe they can rejuvenate you too.

If none of that worked, it was time for the big guns: rubbing some ground hippomanes on your, um, big gun. What's hippomanes, you ask? Well, it's a growth found on the forehead of newborn foals, so just find yourself one of those, harvest it, and then rub it on your penis. Problem solved!

Hey, everybody and more specifically everybody with a penis? Could we give crocodiles a break for like a second? I'm very sorry about the problems you're having with your penis, but if we could just chill on rending crocodiles into component parts for penis medicine I think it'd be a huge boon for human-reptile relations.

WHEN IN ROME

The Romans took more of the an-ounce-of-prevention-is-worth-a-pound-of-cure approach to this particular issue, which sounds reasonable enough. Eat right, exercise—and wear a talisman to protect your boner. One particular favorite was made from the right molar of a small crocodile. Again with the crocodiles!

A more in-your-face amulet choice was some form of *fascinum*, a piece of jewelry in the form of a winged penis worn to honor Fascinus, the "divine phallus god." In fact, Romans were so concerned with this particular ailment that they didn't just rely on one god for help. They also turned to the Greek god Priapus, who started as a minor fertility god with one notable attribute—an enormous, always-hard penis. Romans apparently liked the look of this chap, and he became a very prominent figure in Roman art, literature, and of course, prayers from the afflicted.

Our buddy Pliny the Elder had a lot to say about most diseases and dysfunctions, but he definitely outdid himself when it came to erectile issues. For starters, to get things going in the bedroom, he recommended both leeks and turpentine as aphrodisiacs.

After that, you could move onto some wine flavored with garlic and coriander as the evening progresses, or perhaps a little water that had been used to boil asparagus if you want to get really amorous. If these simpler solutions don't do the trick, never fear: Pliny had a whole list of herbal cures that you might want to try, such as donax, clematis, and xiphium root with pearl barley and wine, or a lozenge made of skink muzzle and feet, mixed with some rocket seed in white wine.

No, we don't know what some of those herbs are, either, but the wine seems to be the key ingredient. Pliny may not have known how to fix erectile dysfunction, but he was pretty darn sure wine was part of the equation.

GREAT PENISES OF THE MIDDLE AGES

Thirteenth-century Bishop Albertus Magnus was many things: a scientist, philosopher, astrologer, theologian, spiritual writer, diplomat, eventual saint—and penis expert. At least, we know that in his treatise *De Animalbus,* he wrote: "If a wolf's penis is roasted in an oven, cut into small pieces, and a small portion of this is chewed, the consumer will experience an immediate yen for sexual intercourse."

But who has time to hunt a wolf for its penis meat when you're busy worrying about your own? Luckily, the good bishop had a backup suggestion: Try sparrow meat. Problem solved!

If all else failed, Magnus suggested eating a starfish, but with a word of warning: it might cause you to ejaculate blood. Luckily, he had a cure for this too: nice, cool lettuce. No wonder the Church considers him to be one of the thirty-six honored "Doctors of the Church."

What if you weren't so into eating and drinking weird stuff? One unlikely, but not unpopular, treatment was to simply surround yourself with beautiful men and women.

If the Hef strategy doesn't pan out, we got one more for you: peeing through your wife's wedding ring, or the keyhole of the church where you were married. Unless you have one of history's chillest priests presiding over your congregation, you'll want to get permission for that one first.

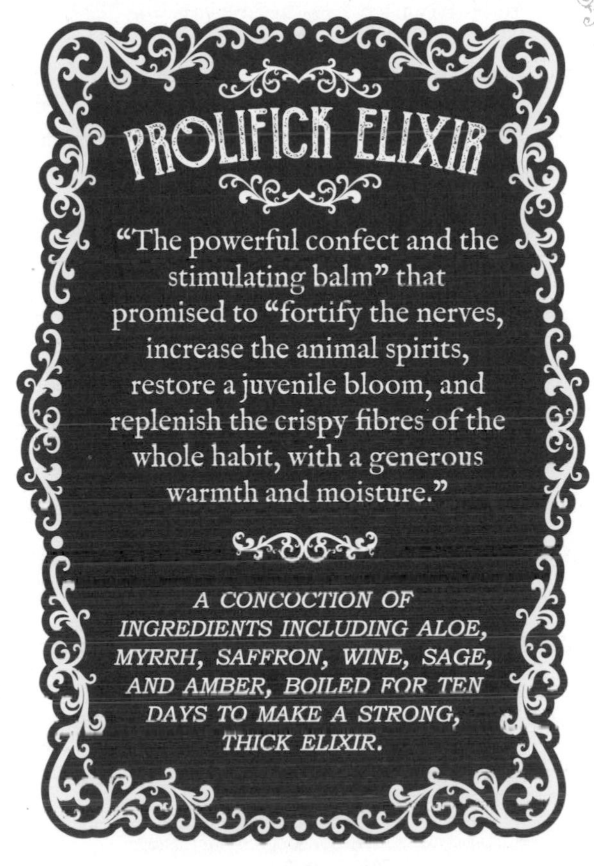

THE ERECTILE REVOLUTION

With the dawn of the Enlightenment came the idea of searching for (or, okay, making up) logical

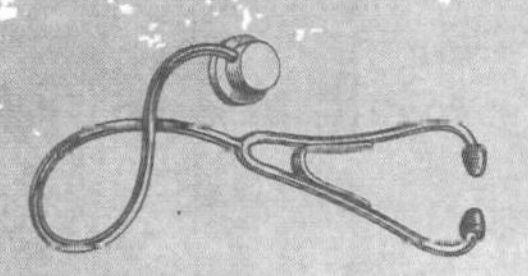

Sydnee's Medical History Corner

At the time that Magnus was writing, it was common to categorize diseases and other conditions by their various perceived temperature and moisture—hot or cold, wet and or dry. Treatments, then, were tailored to address and theoretically correct these perceived imbalances.

In this particular instance, sparrow meat was considered a hot and dry treatment, so it would give you the boost you needed to achieve an erection. A word of warning: It was also known to cause constipation. And really, no one should have to choose between erections and bowel movements.

causes for maladies. For example a pamphlet published in 1783, and titled *The Lady's Physical Directory*, also known as *A Rational Account of the Natural Weaknesses of Women and of the Secret Distempers Peculiarly Incident to Them,* gave many possible explanations for erectile dysfunction.

I just want to take a brief moment to make my distaste for the title of this pamphlet known. It really sucks. Thank you; I'm done now.

The anonymous author (credited only as "A Physician") theorized that the problem at hand could result from either a deficiency of "animal spirits" or those spirits not flowing to the organs of generation (read, "weiner;" it was actually illegal to say "weiner" until sometime in the 1970s, so they resorted to old-timey code words). Stress, excessive drink, and fast living were usually blamed for this.

Alternately, the culprit could be a lack of "animaculae." That term was used to refer to any weird little moving critters that newfangled microscopes had revealed to the world. So, basically, sperm—although we didn't yet really understand how it worked.

Whatever the reason, it didn't matter; the author had a solution for you.

Since this wonder drug was advertised in a pamphlet aimed specifically at women, we're assuming this meant either a lot of uncomfortable dinner conversations, or a bunch of wives secretly slipping a nice dose of penis potion into hubby's after-dinner drink.

FORCES OF NATURE

In the late 1700s, erectile dysfunction was widely thought to be caused by excessive sexual activity, whether with a partner or all by your lonesome. And since electricity and magnetism were popular treatments for just about everything at the time, of course enterprising doctors were going to try them out on penis problems. The field of sexology was more or less founded in 1780 by Dr. James Graham and his Temple of Health wherein—aided by some sexy assistants billed as Goddesses of Health—he treated patients with musical therapy, "pneumatic chemistry," and of course, electromagnetism. He advised patients that they could reverse the erection-killing effects of masturbation and marital excess with cold baths, sexual moderation, and use of his Celestial Bed. Which, we just have to take a moment to talk about. A whopping twelve-by-nine feet, the bed was surrounded by forty glass pillars, and covered by a dome that contained clockwork devices, fresh flowers, and a pair of live turtle doves. A mechanism inside the dome released stimulating fragrances and "aethereal" gases, while organ pipes emitted "celestial sounds." The head of the bed was an electrified clockwork tableau celebrating Hymen, the god of marriage.

In your face, Magic Fingers!

A SHOCKING TREATMENT

Much less fun-sounding was a Dr. John Caldwell, who treated patients with direct application of electricity to the affected body parts or immersion in a bathtub with electrodes.

In 1883, Dr. William Hammond's book, *Sexual Impotence in the Male*, advised attaching electrodes to the patient's spine, perineum, testicles, and penis. A master of understatement, he did say that the effect was "rather unpleasant." Other unpleasant ideas from Dr. Hammond included urtication, a fact term for flagellation, in this case of the buttocks—with stinging nettles.

More of a DIY type? Good news! The Harness

"... You know what, doc? On second thought, I don't really mind my penis just kind of hanging there. I'm just going to learn to live with it, thanks for your time. Did you see my pants anywhere? You know what? Keep the pants, I'm just gonna jump through this window."

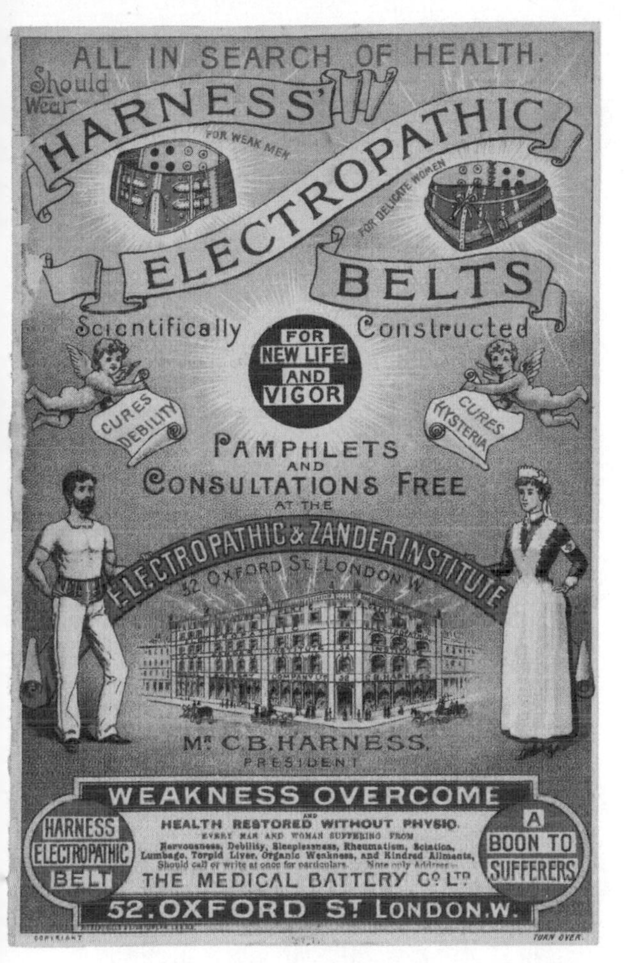

Electropathic Battery Belt, released in 1893, could help with what was tastefully described as "medical electricity for self application."

PUMP IT UP

In the 1840s, we see the first somewhat effective device, French physician Vincent Marie Mondat's horrifyingly named "Congestor," a first-gen penis pump. You put the appendage in question into a glass tube, and a vacuum pump drew blood into the area. A fantastic cure if you just want to display it—not, you know, use it for anything.

The whole pump concept continued to be refined in the early modern era, with such notable items as the Vital Power Vacuum Massager available for sale in the 1920s.

This era also brought us penile splints, which is just what you're imagining—exactly like a splint for your broken arm or leg except they . . . well, except that broken arms aren't typically splinted for the express purpose of inserting them into anyone's orifices. Other awesomely named devices of the era included the Erector-Sleigh and the Virility Cylinder.

DR. LOEWENSTEIN'S HORRIFYING APPARATUS

But no discussion of terrifying love-making technology would be complete without a shout-out to Dr. Joseph Loewenstein's Coitus Training Apparatus. This device was a pair of rings that were connected by wires, and insulated with rubber. To use, you'd position a ring on each end of the penis with the wires stretched between, and cover it all with a condom. You might think of it as training wheels for the penis! (Now, try to think of literally anything else. No luck?)

Loewenstein seemed to think that injury-free use of his contraption was just a matter of practice. He promised that the partner of the "dexterous man" would never even know it was there. His theory was that over time, the penis would remember how to do sex and you could lose the training wheels. Unfortunately, this was never very popular among women who sometimes faced difficulty when it came time to "extricate the apparatus."

That probably didn't work but wowsa, does it ever sound better than electrifying my balls.

PUT THAT THING DOWN!

Excessive masturbation was still the number one suspected culprit in erectile dysfunction, so a lot of treatments were basically medicines intended to restore vitality that had been . . . used up. Some, like Dr. Brodum's Nervous Cordial and Botanical Syrup, relied on relatively innocuous herbs (cardamom, gentian, and colombo) to get men ready for the "married state." Or you could kick it up a notch with Samuel Soloman's Cordial Balm of Gilead, which (of course) had to be taken while bathing your testicles in cold water, or a mixture of alcohol and vinegar. This was also a cardamom-based concoction, but at least this time they threw in some brandy for kicks. Pliny would approve.

Throughout the 1800s, other doctors advised ginseng, strychnine, damiana, yohimbe, and hemlock (some of which appear in dodgy herbal "virility" pills to this very day). You didn't have to find all those ingredients on your own, though. You could simply go to the drugstore and pick up William Acton's strychnine, phosphoric acid, and orange peel combo, or perhaps W. Frank Glenn's special damiana, zinc, arsenious acid, and cocaine mix. (Note: Not available at your local GNC or truck stop. Probably). Other suggestions included massages, drinking urine, or simply refraining from riding a bike. And physician and sex educator Frederick Hollick thought all this searching for a cure for erectile dysfunction was ridiculous because we already had something that made us pretty warm and cheerful and in the mood: Cannabis.

HE'S GOT BALLS (GOAT BALLS, THAT IS)

By the early 20th century, we had finally resorted to surgery. In 1913, a doctor at Northwestern named Victor Lespinasse transplanted slices of human testicle into a man with erectile dysfunctional and claimed it worked so well that four days later, the man demanded to leave the hospital in order to go sate his desires. One year later, Dr. G Frank Lydston implanted a dead man's testicle into his own scrotum. Injections of goat, ram, boar, and deer testicles have also been tried throughout the years. We've even tried implanting chimpanzee testicles in some lucky fellows. (Maybe not so lucky for the chimps.)

BONERS OF TODAY

As we move into the modern era, medical science begins to transition away from these whimsical and/or terrifying solutions into the much duller world of "devices that might actually work." Some of the first, developed in the 1970s and '80s included penile rods and prosthetic devices that could inflate the penis. Injections or suppositories actually placed into the penis are still used in some cases today, as well as penis pumps, tension rings, and implants.

As for magical tinctures, anyone who has watched basically any TV ads in the past decade has almost certainly guessed how this story ends: namely, with prescription medications like Viagra, Cialis, and Levitra. They all work by relaxing the muscles in the penis and increasing blood flow (no electrodes or arsenic required).

WHAT HAVE WE LEARNED?

While the medications dominate the market, there are still other options in use. Testosterone replacement is increasingly popular for a diverse number of medical complaints, including erectile dysfunction, and patients are also advised to try lifestyle changes such as more exercise, less alcohol or illicit drug use, and management of chronic medical conditions.

It may seem like the story ends there, but if the history of medicine has taught us nothing else, it's this: As long as there are people on Earth, and a good number of them have penises, the human race will never stop trying to find new and better ways to fix those penises. As a task it is both hard and lengthy and, yes, that is in fact what she said.

B
A
C
D
F
E
G
H

So What's the Deal With:

SPONTANEOUS HUMAN COMBUSTION?

Sometimes, people just burst into flames for no good reason. Or so the theory goes.

When Did This Become a Thing?

You've probably noticed that a lot of the topics we've covered have their origins, or at least first recorded history, in the ancient world—Greece, Rome, or even Egypt. Not this one. In fact, the first incident of spontaneous human combustion seems to occurred the 17th century. In 1663, Danish anatomist Thomas Bartholin wrote about the case of a woman in Paris who "went up in ashes and smoke" in her sleep, leaving her mattress mysteriously unmarked by fire. (Okay, there was also a guy in the 1400s who supposedly drank so much wine that he vomited fire. Radical, but not spontaneous combustion.)

There've only been a few hundred reported cases of spontaneous human combustion in the intervening centuries, many of them in very similar circumstances. The victim is often an alcoholic, and usually limited in mobility by age, weight, or illness.

Victims are often found near a fireplace or in a room with candles—which may have you wondering what exactly is so mysterious about this. The unusual part is that the fire often consumes only the body's trunk, leaving the arms, legs, and head lying awkwardly around a pile of ashes in an otherwise spotless room.

But People Totally Figured It Out, Right?

Fine question, but if we had cracked spontaneous combustion, you probably would have seen that on the news. We *have* cooked up a number of theories over the years. The fact that many of those who supposedly combusted had been heavy drinkers led to speculation that a lifetime of alcoholism could be the key factor. This idea was widely accepted in the Victorian era, particularly after the mid-1800s when Charles Dickens, at the height of his fame as a novelist, consigned the alcoholic character of Krook in his wildly popular *Bleak House* to this fiery fate.

Others have speculated that intestinal methane could somehow ignite in the body. That's right, kids: Lighting your farts is funny, but it might just be fatal, too.

Kooky psychological theories enjoyed their brief moment in the sun in the 1970s, legitimizing treatment options such as primal scream therapy, or taking hallucinogens and swimming with dolphins. The hypothesis that a person could become so deeply engulfed in depression that they actually became literally engulfed in flames seemed fairly tame by comparison.

Even today, the occasional news story surfaces, as in 2015 when the *Times* of India reported the

death of a baby who, according to his parents, had burst into flames several times in his short life.

Okay, So What's the Deal?

Modern medicine's explanation might seem kind of boring at first, but don't worry. Like most interesting things in science, it gets really weird and gross. The boring part is that as you might have suspected: Passing out drunk next to a fire or dozing off with a cigarette can lead to catching on fire. No methane needed.

So, why didn't the fire scorch all those mattresses and chairs, let alone the victims' limbs and sometimes heads? There's a phenomenon known as the wick effect (sorry, nothing to do with Keanu) that explains it. A candle, burning ember, or lit cigarette may burn through clothing, making only a small hole and remaining hot enough to burn through a person's flesh and begin melting their body fat. Contained by the victim's clothing, the burning fat incinerates everything inside without damaging surrounding objects. The fat fire may burn itself out before spreading to the extremities. As one writer put it, the victim is turned into an inside-out candle. Aren't you glad you asked?

FUN FACT: This phenomenon is generally referred to as spontaneous human combustion, but animals can get in on the fun too. A dead beached whale, for example, can fill with methane as it rots, and combust rather dramatically. A few species of ant in Asia defend their colonies by, basically, blowing themselves up (although, to be fair, not actually bursting into flame). It's more of an extremely dramatic "ant-splosion."

THE DOCTOR IS IN

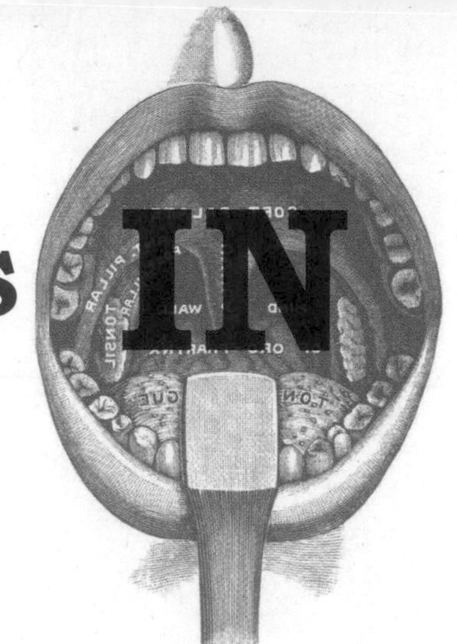

If there's one thing we've learned over the years, besides to stay away from Pliny the Elder's home cures, is that our listeners are a neverending source of weird and wonderful questions. Let's take a break from making fun of our ancestors' crazy ideas, and remember that science is awesome.

Why do people say "an apple a day keeps the doctor away"?

Sydnee: The original saying comes to us from Wales, right around the 1860s. More specifically: "Eat an apple on going to bed, and you'll keep the doctor from earning his bread." Our briefer version first appeared in 1922. It was based on a common belief that fruits and veggies were healthy, which is—refreshingly for 19th-century medicine—not wrong.

Whether or not an apple a day will specifically work to keep you healthy is more complicated; several studies have been done with conflicting results. In 2011, one found that it lowered cholesterol, while another found that Golden Delicious apples may raise it. Another found that eating apples with pears might prevent strokes. However, a study in 2015 found no decrease in doctors' visits with increased apple consumption.

Justin: So what, you're just gonna blow right past the fact that your medical predecessors actually got one semi-right for once? Well not me, dearly departed doctors. I honor you and your only slightly misguided passion for apples.

Why do we put iodine in salt?

Sydnee: We have sort of known we need iodine since ancient times. Chinese writings from 3600 BCE, for example, note that eating seaweed can prevent goiter. If you don't get enough iodine in your diet, you can't make thyroid hormone. If your thyroid isn't pumping out hormones, your brain releases more TSH (thyroid stimulating hormone) to try and, well, stimulate it. Over time, thyroid tissue can get enlarged and form a goiter.

We've got plenty of options regarding consumable iodine sources. It's in seafood and seaweed, and some (not all) salt is naturally iodized. But when we crafty humans realized how important iodine is and that the amount of it present in different food varies, we started adding it to salt. In some places it is added to bread. The first iodized salt appeared in Michigan in 1924; within ten years, incidence of goiter had dropped by around 75 percent.

Justin: Salty stuff should really try advertising its goiter-fighting properties. What? Pringles? No, no, these are actually goiter crisps. And they're prescription.

When an organ is removed, what fills in the empty spot?

Sydnee: Things do shift a bit when an organ is removed. Not migrating or anything, just settling. In areas like the skull or thorax, fluid may fill the space, followed by fibrous tissue.

Justin: Organs need to be more enterprising! If I was a pancreas and I saw a kidney had been removed, you KNOW I'd be stretching my legs out and getting a little comfier. My weird, spindly pancreas legs.

Is popping pimples really that bad?

Sydnee: You'll usually just cause more inflammation, and perhaps introduce more bacteria from your dirty hands. When zits come to a head, you'll probably deroof them when you wash your face. It's the squeezing of zits that aren't ready (especially with dirty hands) that's the real problem.

Justin: Yo, my wife just told you that you had dirty hands *twice*. After you bought her book! Are you just gonna take it?

Why do hospitals say to not eat before surgeries?

Sydnee: If there is one complaint I receive from patients in the hospital more than any other, it's on this topic. it's worth it, I promise.

Many patients receive some sort of general anesthesia during a surgical procedure. Even if it isn't planned, it may become necessary during the course of the surgery. In that case, while the patient is paralyzed or possibly intubated and a machine is breathing for them, they are at a higher risk of aspiration. This means that substances from their stomach could come up their esophagus and go back down their tracheas into their lungs. This can cause major damage and complications. So, in order to minimize this risk as much as possible, we like to keep your tummies nice and empty before a surgical procedure. Obviously, we might take the risk in the case of an emergency.

Justin: Hypothetically, would really wanting to eat a Hazelnut Snickers from this gift basket count as an emergency?

Sydnee: No, I mean emergency surgery.

Justin: But what if I really like Hazelnut Snickers? Hypothetically.

Sydnee: I think we're done here.

Justin: Hypothetically?

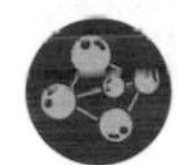

Can I eat moldy food if I just remove the moldy part?

Sydnee: I wouldn't. There could be single fungal spores that you couldn't see. Individual spores are microscopic and they could have permeated the food. Would that few number of spores kill you? Probably not, but why take the risk?

Justin: Listen, I love English muffins as much as the next guy, but no English muffin is worth risking death . . . okay, I've had a few English muffins that would be worth risking death, but they're few and far between.

Why does my stomach growl when I'm hungry?

Sydnee: Technically, that's not your stomach, it's your small intestines. Borborygmy is a gurgling noise made by fluids and gas moving around in your intestines. It doesn't just occur when you're hungry, though; borborygmy is also a perfectly natural result of typical digestion.

Justin: And if your *Justin* is grumbling, it probably meant he had to go to the DMV.

TREPANATION

You need this procedure like you need a hole in the head. Now, just stay still while we get the drill lined up . . .

Wherein we learn that drilling a hole in one's own head as an attempt to treat disease is actually a pretty bad idea, believe it or not—except when it isn't, which is an even stranger thought. But when you're down to that option, it's likely to be appealing anyway, compared to worse alternatives . . . right?

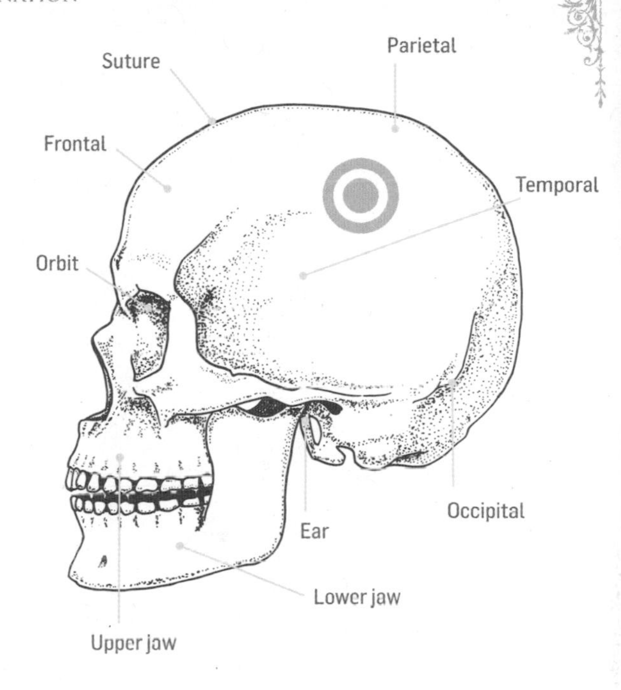

We try to take it pretty easy on those ancient medical practitioners here at *Sawbones*, broadly speaking. Sure, we'll give them some razzing. A bit of gentle ribbing, perhaps. But we understand that many of them were just doing their best and really wanted to help people.

But when our misguided predecessors start drilling holes in each others' heads, you might assume we'd dispense with that courtesy. That, well, that's just wrong, isn't it? That's where your brain is, right? Shouldn't we be keeping it intact?

The shocking thing, as it turns out, isn't that the ancients were making holes in perfectly good skulls. It's that some of them were saving lives.

A LITTLE OFF THE TOP

"Trepanation" is derived from the Greek word *trypanon* which means "to bore," which is a funny choice for what has to be one of history's most "you'll pay for your whole seat but you'll only need the edge" thrill-level procedures, especially in the pre-anesthesia days. And we're heading back to the pre-pre-pre-anesthesia era.

The oldest trepanned skull was found in a neolithic burial site in Ensisheim, France, which would place the practice as being at least 7,000 years old. Now, if we had only found the one, we could have chalked it up to the misguided surgical efforts of an ancient quack. But this was not an isolated incident. Despite the fact that drilling a hole in your head is usually, wowza, just a super bad idea, it was practiced throughout the ancient world in Egypt, China, Greece, India, Rome, and in early American civilizations.

What could it fix? Well "being too good at math," for starters. But the ancient cultures that practiced it had numerous and varied reasons for doing so. They believed it could fix everything from headaches to epilepsy, anxiety to evil spirits.

SO HOW'D THEY DO IT?

Let's say you wanna drill a hole in your head, but you don't want a bunch of newfangled technology getting in the way. You want a hand-drilled

JUSTIN VS SYDNEE

JUSTIN If you in your day-to day life are ever told that the treatment you're employing is also useful for dismissing evil spirits, here's a good life hack: Throw a chair through the nearest window, attach a grappling hook to the ledge and rappel right the heck out.

SYDNEE No, actually, please don't do that. That sounds very dangerous.

JUSTIN It's fine Syd, just having a little fun. Reader gets it, we've been kidding around like this. It's kind of our thing.

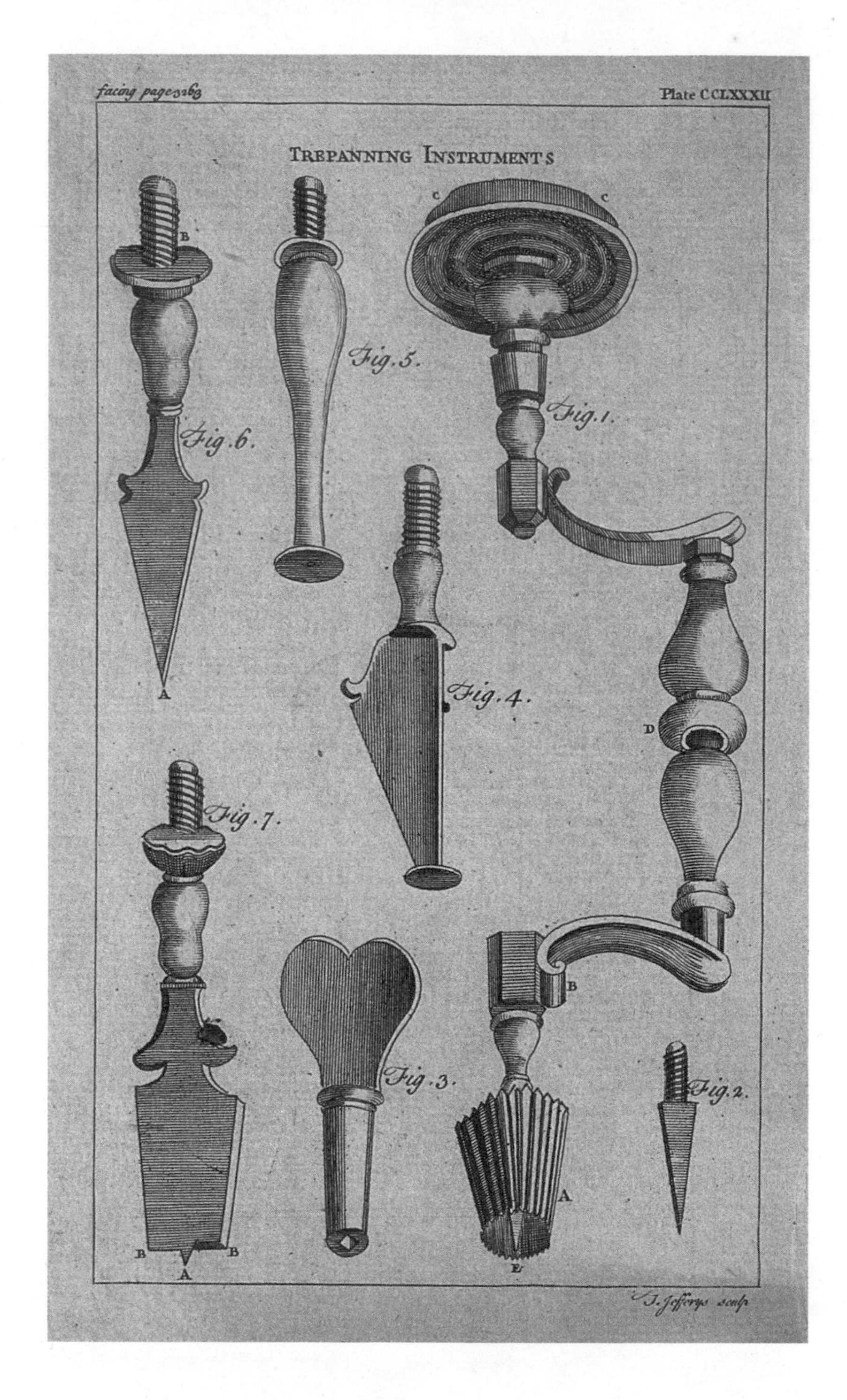
facing page 3263
Plate CCLXXXII
TREPANNING INSTRUMENTS
Fig. 1.
Fig. 2.
Fig. 3.
Fig. 4.
Fig. 5.
Fig. 6.
Fig. 7.
J. Jefferys sculp

LAST CHANCE TO BAIL:

Hey! If you'd prefer not to understand how to destroy a skull on a deep, personal level, this would be an ideal time to bounce. *You can rejoin us on page 64 after the really yucky stuff is over!*

classic, like grandpa used to do. We hear you.

You've got plenty of options with regard to size. The holes the ancients drilled could range from a few centimeters to half the skull (which may not even qualify as a hole anymore at that point, but why quibble?). The most popular real estate for drilling tended to be right at the top through the parietal bone, followed by occasional holes in the occipital or frontal bones. If you wanted to get really fancy, you could even try the temporal region.

Trepanation methods varied throughout history, but all were similar in one key way: We were trying to make a hole in a thick, hard bone designed to keep things out.

Initially, some ancient cultures would use sharp stones, such as flint or obsidian, to sort of scrape the bone away. A tool called a *tumi* was used in Incan cultures for this purpose. It was a ceremonial knife that you would use to create sort of a tic-tac-toe board or hashmark on the skull, and then pry out the square of bone inside. This seems really intense, but don't worry! The surgeon would be bracing your head tightly between his knees while he did this, and he was a professional.

In ancient Greece, trepanation was mainly done for the purpose of relieving pressure on the brain in the case of a skull fracture. Practitioners had a number of tools for the job. Among them was our favorite, the *terebra*, a cross-shaped boring tool that made tiny skull holes in a pattern until you could punch out the bone inside. Yes, a skull

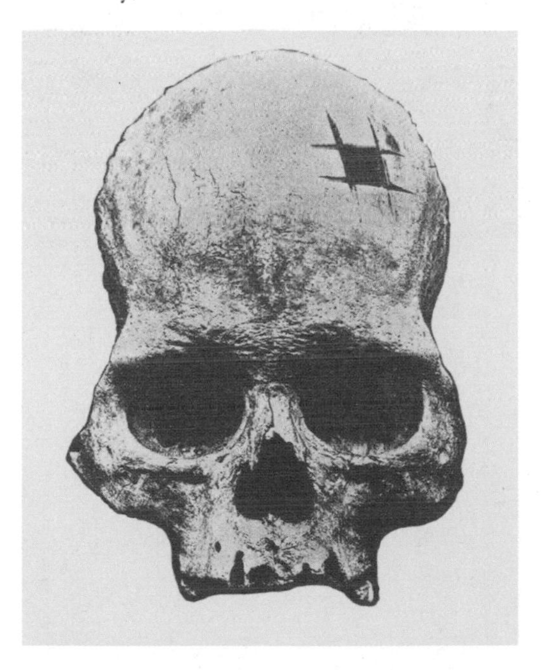

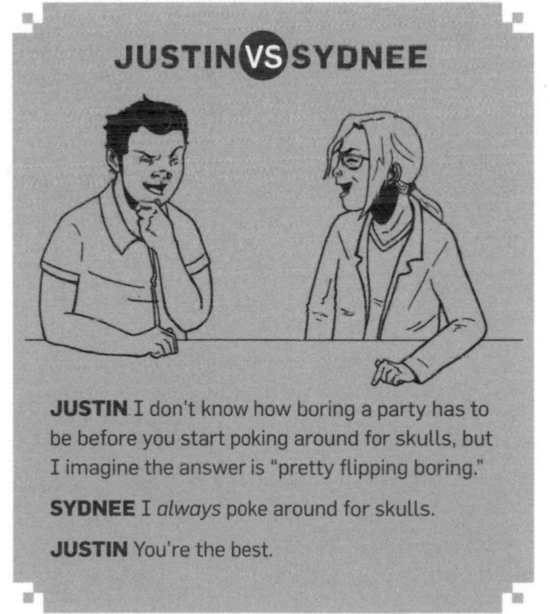

JUSTIN I don't know how boring a party has to be before you start poking around for skulls, but I imagine the answer is "pretty flipping boring."

SYDNEE I *always* poke around for skulls.

JUSTIN You're the best.

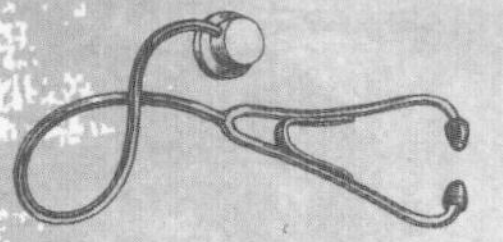

Sydnee's Fun Medical Facts

Every med student knows the name Broca. "Broca's Area" of the brain is the third convolution of the left frontal lobe next to the lateral sulcus and if it is damaged the result is Broca's, or expressive, aphasia. With this condition, someone would be able to understand what they heard, but their own speech would seem stilted and labored.

perforator. Good for trepanation, but also handy for scrapbooking!

By the 12th century, our suite of tools for skull perforation had evolved to a point where circular cuts, rather than jagged hashes, were the standard (progress!). The Medieval era brought with it mechanical drills, circular saws, and all kinds of terrifying props lifted straight from Jigsaw's torture den. Even though these things were documented, we seemed to have completely forgotten this history by the 19th century.

JUSTIN VS SYDNEE

JUSTIN That's sad, in a sense. Those drills do look really cool. But, you know Syd, it's probably for the best, considering that, well, we're talking about increasingly creative ways to straight-up murder people.

SYDNEE Well, actually Justin, you might be surprised to learn that trepanation wasn't necessarily the death sentence you are assuming.

JUSTIN You're joking.

SYDNEE Shh; don't spoil the next section.

CAN'T WRAP MY HEAD AROUND THE IDEA

Let's say you gathered up all the trepanned skulls that researchers had found over the years. After you were either arrested or made the subject of a reality TV show, you'd notice something interesting. By charting new bone growth after the procedure (you know how to do that in this hypothetical) as well as the presence of multiple holes, you could infer a trepanation survival rate of around 60 percent. You might be tempted to scoff at that, oh so snooty 21st-century dweller, but maybe first go make a big hole in your head and let us know how it works out. (Note, if you're perusing this in a bookstore, please pay for it and exit the store first.)

Drilling a hole in one's head is so unintuitive, in fact, that it took modern folks a little while to believe it could even be done (without the patient's think-meat gooshing out of their head, that is).

The first evidence of this practice in days gone by—a trepanned skull, naturally—was found in 1685 by Bernard de Montfauchon in France, and subsequently disregarded. In 1816, another skull turned up, also in France, and it was understood that a craniotomy (that's drilling in the skull) had occurred. But the medical community still assumed the procedure had happened after death or . . . you know . . . during.

JUSTIN VS SYDNEE

SYDNEE We're saying that they died as the result of a traumatic, penetrating head wound and open skull fracture.

JUSTIN Yes, sweetheart, I think they get it.

HEY, WHAT A PARTY FAVOR!

All of this speculation changed as a result of a party in Peru in the 1860s. Ephraim George Squier, an archaeologist, journalist, and U.S. Commissioner to Peru, made the key discovery at a party held by the very wealthy Señora Zentino. She had an amazing collection of pre-Colombian art and artifacts, and among these antiquities, he noticed a skull with a hole in it.

With Zenito's permission, he took the skull to the United States, and presented it to the New York Academy of Medicine. While many agreed that the skull did seem to present evidence of a surgery done while the patient was still alive, a vocal minority still argued that the injury had to be postmortem. There was simply no way that a patient could have survived this procedure. Surely, even primitive societies would have known better than to attempt it. Something everyone agreed on: No ancient, nonwhite civilization could have beaten the hyper-advanced modern white man to such a complex surgical feat. Ah, racism, the great uniter.

BROCA JUMPS TO THE HEAD OF THE CLASS

With a hearty "I'm outta here, and I'm takin' my skull with me!" shouted over his shoulder, Squier then crossed the Atlantic to Pierre Paul Broca—physician, anatomist, and founder of the Anthropological Society of Paris. One hiccup: Broca was also one of the main proponents of the racist ideas that had kept others from accepting the validity of Squier's theory. But in this case, Broca let the science do the talking and agreed that the skull did seem to represent an early form of trepanation.

"What astonishes me is not the boldness of the operation," Broca said at the time, "as ignorance is often the mother of boldness." Just couldn't leave well enough alone, could you Paul?

Broca published his and Squier's findings and, in doing so, stoked the medical community's interest in trepanation. The hole-having-head hunt was on, and by 1867, many more crania had been discovered in France and Peru. After collecting loads of skulls (and presumably staging some pretty radical haunted houses), medical historians finally reached a consensus. Not only were these holes made intentionally while the patient was still alive, but that this practice crossed geographical and cultural barriers throughout history.

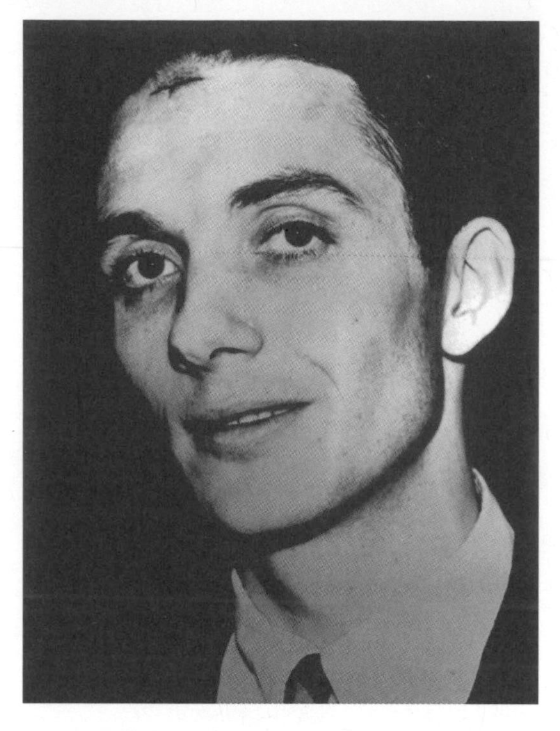

TREPANATION FOR FUN AND PROFIT

In the 1960s, a Dutchman named Bart Huges who had attempted (but not completed) medical school, learned about the ancient procedure and became obsessed with it. In lieu of any actual evidence as to why trepanation had occurred, he began to theorize that it was a way to regain the joy of youth. Why? Well, babies have fontanelles or "soft spots," open areas between the bones in their skull to allow for growth.

JUSTIN I think it's pretty messed up that babies come with a self-destruct button, but Sydnee says I shouldn't call it that.

SYDNEE I'll grant you, they can be kind of unnerving, but don't worry. These holes close naturally as we age.

Huges believed that children are so happy and open-minded (literally and figuratively) thanks to these spots and, by comparison, the skulls of adults are just too darn restrictive. If we could just recreate those holes in our heads, he thought, we would be able to reconnect with our joyful past. We could loosen our ties, spread open our skulls, and just be.

His impressively odd, yet medical-sounding, explanation for this phenomenon hinged on a concept he called "bloodbrainvolume." Actually, it's better if we let Huges explain:

"I met [someone who] used to stand on his head . . . for considerable periods of time. When I asked him why he did it, he said it got him high. [Later, I was given] some mescaline, and it was then that I got my first clear picture of the mechanism, realizing that it was the increase in the volume of brainblood [sic] that gave the expanded consciousness . . . [which] must have been caused by more blood in the brain which meant there must have been less of something else. Then I

realized that it must be the volume of cerebrospinal fluid that was decreased."

(A side note about Huges which should be zero percent surprising at his point: He was also convinced that psychedelic drugs could achieve these same effects, but only temporarily. He called drugs like LSD, mescaline, and marijuana "psychovitamins." He also named his daughter Maria Juana.)

Sure, he could have just made a hole in the bottom of his spine to let out the extra fluid, but that wound would heal, so it would only be a temporary fix. Huges realized he could relieve the pressure with a hole in the skull, and those things never heal, so he'd be golden!

He published his theory as both *The Mechanism of BloodBrainVolume (BBV)* and *Homo Sapiens Correctus* in 1964. Also of note: he published it on a scroll, which made it awful hard to shelve in your average medical library.

THE SEARCH FOR AN OPEN-MINDED DOCTOR

For two years, Huges traveled around looking for a surgeon willing to help him prove his theory, but none agreed. So in 1965, armed with only the completely made up theory of bloodbrainvolume, Huges drilled a hole in his head, using an electric drill and a scalpel. (In hindsight, we should have listed those items among the things Huges was armed with.) He revealed the fruits of his roughly forty-five minutes of labor to an awestruck, and probably grossed-out, crowd in Amsterdam as an art happening. The "art" part is debatable, however, it's hard to argue that something most definitely did happen there.

His next trip was to the local hospital for an X-ray of his new "third eye," but once the doctors saw what he had done, he was detained in the psychiatric ward for three weeks.

While few were inspired to copy his actions, the two followers he gained were all-in. Joseph Mellen and his wife Amanda Fielding both sought their own at-home, Huges-inspired neurosurgical procedures. Mellen wrote an autobiography centered around his experience called *Bore Hole.* Not to be outdone, a short film was made of Fielding's trepanation adventure called *Heartbeat and the Brain.* (Presumably because *Dumb and Dumberer* was taken.)

A self-styled trepanation evangelist, Fielding attempted to bring the procedure to the mainstream by running for British Parliament (twice). She only got a total of 188 votes, but that's not a bad showing, considering the entirety of her platform was "free trepanation for all."

Another fan of Huges, Peter Halvorson, went on to form the International Trepanation Advocacy Group, ITAG. This group still exists today, working to find surgeons who will perform elective trepanation and research the effects that this procedure has in terms of blood flow and volume and cerebral function. As of yet, not much evidence exists, but honestly, who'd show up for that clinical trial.

DO WE STILL DO THIS?

We do have legitimate reasons to drill holes in the skull in modern medicine, but they have nothing to do with bloodbrainvolume, because that is not a real thing. Nothing to do with evil spirits either, for that matter. In case you were wondering.

But it is a fact that in cases of intracranial bleeding, pressure on the brain tissue can result in damage. In emergencies, the treatment to preserve brain function is to drill a hole to relieve pressure or remove a piece of the skull temporarily. During brain surgery, it is also necessary to remove a piece of the skull and this must be left out at times due to post-op swelling. It can be kept safe in . . . you'll never guess where.

Go on, guess. Nope? The abdomen. Aren't bodies the coolest?

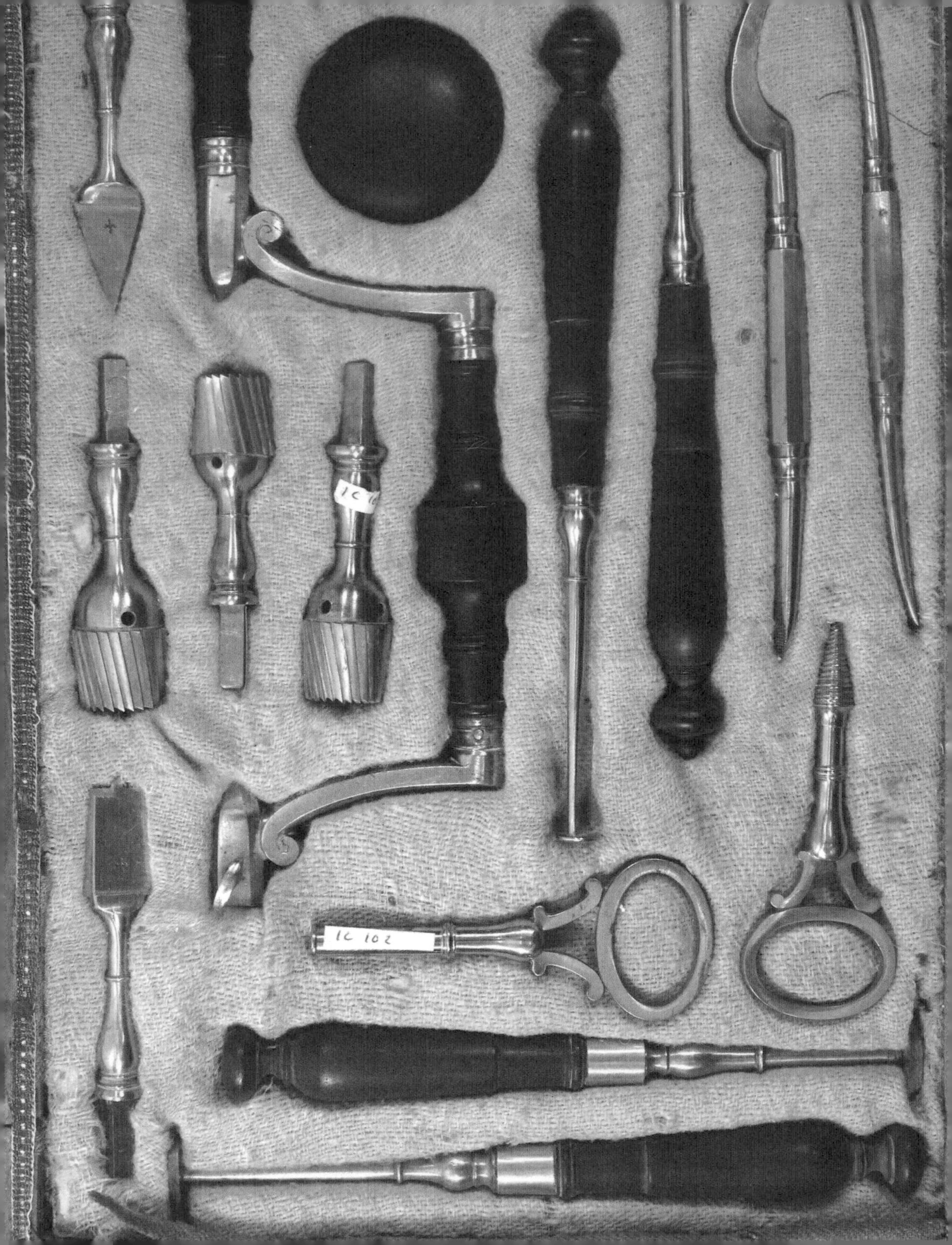
IC 102

SURPRISE POP QUIZ

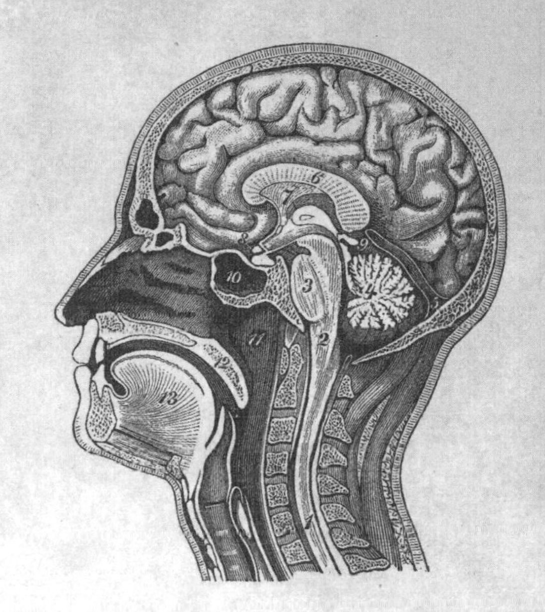

Hey kids! Can you select all of the strange things that head-drilling enthusiast Bart Huges actually said in interviews? Don't be fooled by the ones that seem just too bananas to be the genuine article—this partisan of trepanation carried around some very, shall we say, interesting thoughts in his head.

- ◯ "With enough blood, the central nervous system is a better doctor than your doctor."
- ◯ "Eat a salad every day."
- ◯ "Gravity is the enemy. The adult is its victim—society is its disease."
- ◯ "My problem is how to explain to the adult that he has too little blood in his brain to understand, if he has too little blood in his brain to understand that."
- ◯ "I suppose in cases of severe adulthood, there might be a depression immediately after the operation in a period of retrospection."
- ◯ "I think it's a good idea to exchange the unnecessary words for colors, and keep the few left for communication of information."
- ◯ All of the above. And more.

ANSWER KEY: All of them. He said all of those words. In that order. Also, he had a point about how awesome salad is.

THE GROSS

No, you read that right: The last section wasn't the gross stuff. Strap in!

The banquet's begun, eat some mummy
We hear the poop slushie's quite yummy
Need your humors aligned?
Glass of pee will do fine
Oops, it gushed out the hole in your tummy!

Mummy Medicine and More

Eating People
Is
Bad.

Usually we try to let our narrative unfold naturally and spring the surprise ending on you at . . . well, at the ending. But this time we've put the ending right up front: Don't eat your friends, your neighbors, your fellow parishioners, nobody. It won't help. We promise.

LOVE AT FIRST BITE

You didn't need us to tell you that eating people is generally frowned upon. Sure, we all ended up giving Hannibal Lecter a pass, but when facing the combined charisma of Anthony Hopkins and Mads Mikkelsen, what other choice did we have?

Doctor Lecter aside, there aren't many ethical debates nowadays about whether or not we should engage in cannibalism. There's pretty much a consensus that it's kind of mean, and also basically pointless. However, as with most things that we now realize are completely terrible ideas, humans had to try eating each other for a while before they could really have an informed opinion about it. Besides, our ancestors *had* to take a stab at using one another for medicine—if for no other reason than that they had tried pretty much everything else in some cases.

We're not necessarily (or very often at all) talking about eating whole people . . . just the medicinal bits! The Romans, for example, tried consuming human blood to see if it had any benefits. Specifically, they drank the blood of gladiators as a treatment for epilepsy. When that proved unsuccessful (surprise!), they moved on to eating bits of their livers instead, presumably without fava beans, or even a nice chianti.

GOBBLING GUYS FOR GOD

Cannibalism for religious or spiritual reasons dates way back. The Aztecs sacrificed people to honor their gods and, as part of the ceremony, would also eat some of the sacrificial organs.

Throughout history, various peoples have believed that consuming another person allows one to take on their attributes, such as eating the bodies of enemy warriors to obtain their courage. Cannibalism could also be employed as a punishment when the death penalty just wasn't enough. Some early tribal cultures believed that the soul stayed with the body for three or four days after death. If they ate the body of their enemy within that time period, they would prevent the soul from ever ascending to heaven.

For other cultures, such as the Brazilian Wari, eating the bodies of their own tribespeople was part of their sacred duty to honor and respect their spirit. This continued as late as the 1960s.

CANNIBALISM MAKES A COMEBACK

In the 16th century, the practice of eating people for medicinal purposes came back into vogue in Europe. Basically, some folks looked at the fact that ancient peoples thought cannibalism was beneficial, and wondered whether that was due to some underlying medical reason. Without much actual anatomical understanding, these scholarly types then decided to see whether, in fact, human corpse parts, bones, blood, and the like could be used as medicine.

At this point, you've probably furrowed your brow to a migraine-inducing degree. Why would anyone, no matter how generous their take on ancient tribal practices, make the logical leap of eating their buddy for medicinal purposes? It's not as wild as it sounds. Well, *it is*; let's not lose sight of that—but there are three reasons that this odd behavior started to gain ground.

The first thing to understand is that, back in those times, the idea of "spirit" was an important component of how we understood health. While it could not be seen, the spirit was thought to be physically contained within the body. It stood to reason (sort of) that a good way to maintain your own spirit was to consume someone else's.

Another contributing factor was the popularity of homeopathy at the time. While homeopathic medicine does not inherently suggest eating other people, it does incorporate the idea that "like cures like." In this instance, one may conclude that a good cure for a headache would be to eat someone else's head. Well, their skull to be more precise. A little bit of it, anyway.

The third important notion to understand is that it was commonly believed that all creatures have a predetermined life span—but also that it's possible to accidentally die before you use up those years. If one person died too early, someone else could harvest those extra years by . . . eating them. You know how enemies in *The Legend of Zelda* sometimes drop hearts after Link kills them? It's sort of like that, except you never saw Link frying up a Gannon flank steak.

All that being said, it's unlikely that anyone was eager to commit murder just to give this bizarre treatment a try (and if they were, they weren't foolish enough to leave evidence), and grave robbing was a high-risk occupation. Finding well-preserved dead bodies that no one would mind you eating was quite the challenge. However, the answer arrived when Europeans discovered the newest taste sensation: mummies.

YES, WE'RE ABOUT TO EAT MUMMIES

By the Medieval period, Europeans had been stumbling upon the well-preserved bodies of Egyptian mummies for some time, and digging them up without much plan as to what to do with them. Sure, you could sell them to be showcased as curiosities, you could prop one up in your living room to scare neighborhood kids, but

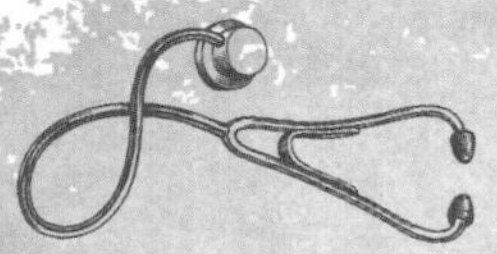

Sydnee's Medical History Corner

This whole mummy-eating thing might have resulted from a wacky misunderstanding. Here's the deal. Medieval Islamic medicine used a pitch-like resin known in Arabic as *mumiya* as something of a cure-all, and the Western world came to use it as well, calling it "mumia." In the 12th century, imported Asian mumia ran low for, and according to contemporary history, merchants started looking around for a likely substitute since that block goo was a real top seller. Turns out Egyptian mummies have some black goo in them, and the words sound so similar that the ensuing confusion in translation resulted in the term "mumia" being applied to the sticky, black substance that could be scraped out of preserved bodies from Egypt. Henceforth, the theory goes, "mumia" was a goo made from mummies, and all of the curative properties previously ascribed to pitch now became associated with ground-up old bodies.

what then? We didn't really start scientifically studying mummies until the 20th century. In the intervening centuries, they were just sort of . . . there. So why not try using them for medicine?

The flesh of mummies was really dry and crumbly; for some reason, this was seen as evidence that they made good medicine. The bodies were dug up, ground into powder, and added to various tinctures. Every apothecary worth his salt would have had a jar of mumia available for sale, and it was prescribed for, well, everything—from gout and epilepsy to bleeding and clotting.

MUMMY-WHILE-U-WAIT

How many mummies did we condemn to this most ignoble of final rests? Well, enough that we developed a serious kink in the mummy supply chain. As mummies were a limited resource, and not everyone had access to a pyramid and a team of archaeologists, the price started going up. Mummies were being sold for up to five pounds British Sterling per pound of mummified flesh, which was a lot of money in the 16th century. (If you see a mummy around for that price these days, buy one. Heck, buy two and send us one.)

Our insatiable desire to eat mummies led some to try an instant-mummification process. At some point we hit upon the idea of small batch, slab-to-table insta-mummies. Men and women who died while still young and strong were placed in mixtures of honey and herbs to preserve them quickly, and try to create some of the medicinal properties inherent to mummies.

Classic rookie mistake: You can't rush great mummy. You can go down to any McMummy's and get one with fries, but for the real benefits, you need grade-A dry-aged mummy, like grandma used to make—speaking of which, Grandma, how are you feeling?

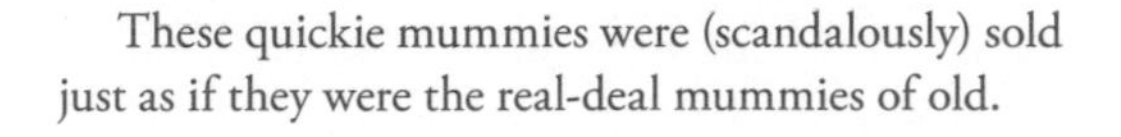

These quickie mummies were (scandalously) sold just as if they were the real-deal mummies of old.

BREAKING IT DOWN: WHAT TO EAT AND WHEN TO EAT IT

So, what illnesses and ailments could medical cannibalism actually treat? In a word: nothing. But what did we *think* it was good for? That varied depending on which part you ate.

MUMMY MEAT

Preserved human flesh was crumbled into various tinctures that were intended to stop bleeding—or, conversely, to break up blood clots. (Note: When it's claimed that a treatment does two opposite things, it's a good time to get suspicious.) The substance was also supposed to help treat a cough, as well as menstrual problems. Mummies were said to be best for this, but any flesh would do equally well (read: not well at all).

FRESH FLESH

If you didn't have a mummy, you could still make at least some medicines from newer donors. One recipe in particular involved obtaining the flesh from the cadaver of a reddish-haired man, age twenty-four (no less, no more), who had died from violence, not illness. The body was to be cut into chunks, combined with myrrh and aloe, and then soaked in wine for several days. This resulted in . . . well, whatever it resulted in, it was supposed to be good for epilepsy. That said, once you've gone to the trouble of finding an unlucky ginger dude to sauté, you may as well try eating him for other stuff.

Another recipe suggested that you pulverize a human heart and take one dram of the stuff on an empty stomach every day to cure . . . dizziness. Hopefully, *really severe* dizziness.

And if you have a nasty bruise, a paste made from corpses will clear it right up.

SKULLS AND BONES

Human skulls were generally a little easier to track down than whole bodies, so there were a lot of uses developed just for the old head bones. Ground-up skulls were used in medicine for both headaches and epilepsy. You could also take that ground-skull powder, mix it with chocolate, and it take it for apoplexy, an outdated blanket term that generally referred to bleeding in the brain or a stroke.

King Charles II was such a fan of using skulls as medicine that he had his very own name-brand medication, King's Drops, made of powdered human skull mixed with alcohol.

(By the way, Charles wasn't the only royal fan of medicinal human remains. In fact, one hundred years earlier, King Francis I of France had been on trend, carrying some mummy meds in a pouch at his waist at all times.)

As with most medications, not all skulls were created equal. If you could find some that had moss growing on them, you had a much more powerful cure on your hands. In fact, the moss that grew over a buried skull could be removed and used as medicine all by itself. There was a specific name for this moss, "usnea," which was supposedly good for epilepsy and nosebleeds.

I don't endorse many products; I have to be careful about my brand. So you know you can trust me when I say that King's Drops are the very best skull-based headache remedy on the market. Now with 50 percent more skulls!

WHEN TO CHEW THE FAT

The uses for human fat were more limited, since it was probably a little harder to get—needing a fresh, well-preserved corpse and all. Still, we do read of people using bandages soaked in human fat for improved wound healing. It was also rubbed into the skin for gout, arthritis, and rheumatism.

LITERALLY BLOOD-THIRSTY

Thanks to *Twilight*, *Interview with the Vampire,* and *Dracula: Dead and Loving It*, the idea of drinking human blood probably doesn't seem that shocking to you. Underneath all those vampire stories are tales of actual humans tossing back actual human blood to see what it would do. The supposed "cures" were derived from the idea that blood gave you more of a life-force of sorts, put a spring in your step, and restored the bloom and blush of youth.

Since blood was thought to be good for vitality, the closer to fresh you could ingest it, the better.

Will someone for the love of *God* please invent coffee?

In addition, the younger, healthier, and (just when you thought this sentence couldn't get any grosser) more virginal the donor, the better.

During executions in the Medieval era, people would stand at the scaffold, offering bribes to the hangman to pass them a cup of fresh blood after it was over. In the case of beheadings, some folks desperate for a little vim and vigor would just stand close enough to the event so that they could open their mouths and . . . um, catch the spray. Or, if that seemed a bit uncouth, maybe get a discreet go-cup and take it home, where you might—according to a 1679 recipe from a Franciscan apothecary—cook up some nice old-fashioned blood marmalade.

This practice also resulted in some poor people selling their blood for money. Sure, today's beer-starved college students still make cash with their plasma, but it likely isn't being sucked directly from their arm by an old rich person. Therapeutic bloodletting was also in vogue at this point, so there was an added incentive for those who did the bleeding to siphon off a little more than was needed. The excess blood for sale was advertised via jars in their shop windows.

Finally, if no one was dead, and you weren't a murderer, and you couldn't afford to buy some—but you really needed to feel young again—you could always just drink menstrual blood. If you wanted. Up to you.

DO WE DO THIS TODAY?

Most all of the modern world has obviously lost its taste for other people, aside from isolated cases. The closest we have come to consuming our flesh medicinally is the recent trend of eating the placenta in order to cope with postpartum depression.

There are some remote tribal cultures in the developing world that still practice cannibalism, but largely for spiritual reasons, as opposed to a purported medicinal benefit.

And no one, thankfully, is eating any mummies. Until someone figures out a way to power iPhones with them, our most precious natural resource is safe.

MIRACULOUS UNIVERSAL CURE-ALL MERCURY

Ladies and gentlemen, let's get to know mercury, your friendly metal next door. Its symbol is Hg; its atomic number is 80. It's been taking ukulele lessons in its spare time, and it's the third heaviest element in nature after gold and platinum (you might say it got the bronze, if that wouldn't be terribly confusing).

It's also the only metal that's a liquid at room temperature, which means it's really fun to play with and also really dangerous to play with—and we . . . we should have lead with that second part first, shouldn't we?

Mercury is toxic to most living things, including humans. Exposure causes a wide variety of symptoms, from difficulty breathing and loss of hand-eye coordination to a "metallic taste in the mouth," which seems self-evident, but okay, sure.

It does seem magical, watching metal goosh around on a table, so it's no surprise that people assumed it had magical healing properties. We rubbed the stuff on everything, hoping that quicksilver would share a bit of its magic with a suffering patient.

Not to spoil the ending, but we eventually figured out that mercury is very bad for you. It took us longer than it maybe should have, because the wide variety of mercury poisoning symptoms were often confused with a relapse of the disease the mercury had been used to treat.

What were those diseases? Oh, friend, get comfy.

USED TO TREAT:

MORTALITY

Ko Hung, a 4th-century Chinese alchemist, made a mercury-based elixir that he believed could help a person live longer or maybe forever. He didn't stop there though. Ko Hung also thought you could rub mercury on your feet and walk on water, or mix it with cranberry juice to help older men become more fertile. He was wild about the stuff.

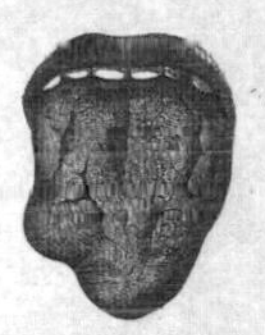

SYPHILIS

We started trying to treat syphilis with mercury way back in the 14th century. It was such a common treatment for a time that it gave birth to the adage "A night with Venus, a lifetime with mercury." There's some historical evidence of syphilis clearing up after mercury treatment, but there's also evidence of syphilis getting better for no apparent reason. Regardless: Don't do this. Mercury is bad for you, remember? We have real medicine now.

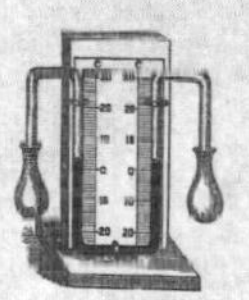

TEMPERATURE

Mercury has been used in glass thermometers since Daniel Gabriel Fahrenheit came up with the technique in the early 1700s. These days, mercury thermometers are rare.

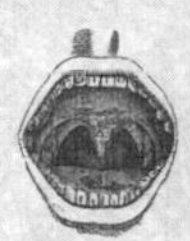

DENTAL CAVITIES

Mercury is one of the components in dental amalgam, a combination of metals that have been used in dentistry to fill cavities for 150 years. The FDA says that dental amalgam is safe, with levels of mercury well below those deemed dangerous for humans, regardless of what various scare stories on the Internet may tell you.

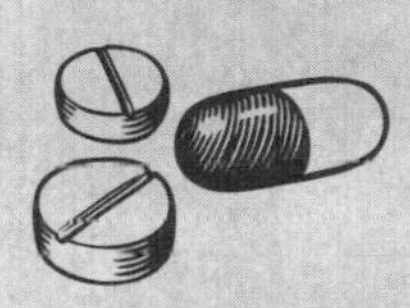

DEPRESSION

In the 1800s, physicians used the term "hypochondriasis" as a sort of blanket diagnosis or at least descriptor for several different disorders. One of them was depression, which was the particular flavor of hypochondriasis that prompted future president Abraham Lincoln to seek treatment. Lincoln started popping little pills called "blue mass," the chief ingredient in which was mercury. There are reports the he behaved erratically at the time, which some historians chalk up to mercury poisoning.

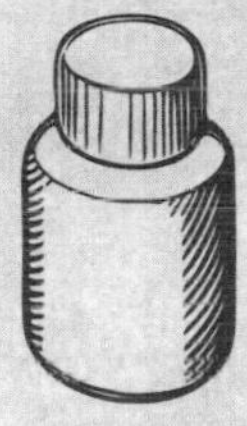

INFECTIONS

There's a whole generation of theater nerds old enough to memorize every word of *Rent* but young enough to have no clue why one of those words is "Mercurochrome." That's because the topical antiseptic, while still available in much of the world, is banned in the United States, and later, in a handful of other countries due to its mercury content . . . well, sort of.

In 1938, the FDA rated a bunch of meds that had been used for years with no ill effects as "Generally Recognized as Safe." Mecurochrome, generically known as "merbromin," was one of them. This meant that, basically, it hadn't been safety-tested, but it seemed pretty harmless. The FDA pulled this designation in 1998, so to have it classified as safe, a corporation would have had to step up and pay for the testing. But by that point, there were a lot of antiseptics that were more profitable and effective, so nobody wanted to pay to test Mecurochrome, and it was sort of banned by default. Poor Mecurochrome.

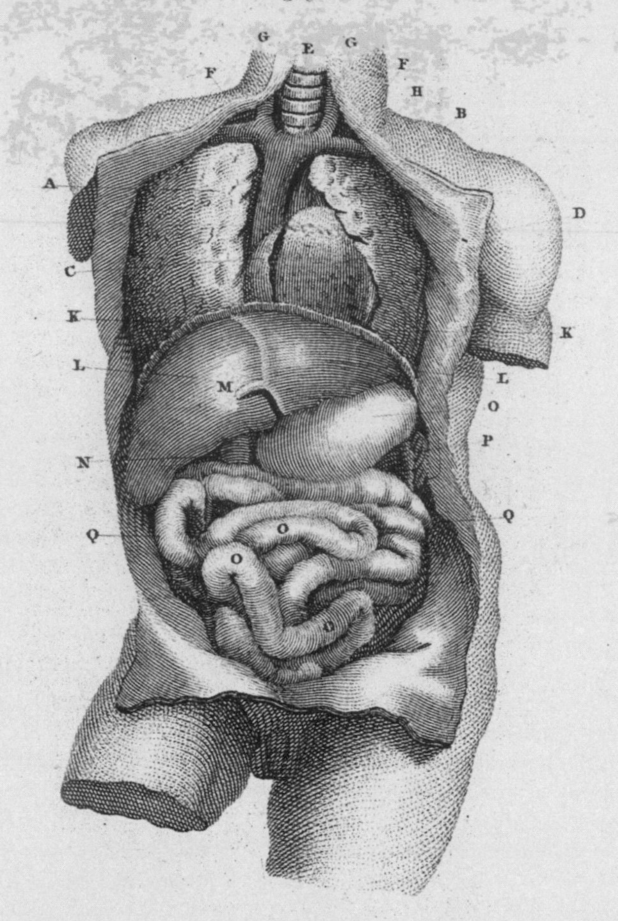

Guthole Bromance

Love can bloom in the most unlikely of environments, even the acidic contents of the human stomach.

Okay, maybe love is overstating it, but this is as close as Sawbones *gets to romance. Introducing Dr. William Beaumont, who jammed his hand into the gaping shotgun wound of a burly fur trader and found within, medical history's most unlikely friendship.*

Prior to the 1800s, the digestive process was a mysterious one, and doctors were divided as to how it worked. One camp believed that it was a purely mechanical process, in which food was tossed about in the stomach and intestines, gradually getting mushed into smaller and smaller bits until it was unrecognizable and drained of all nutrients. Others believed that chemicals were involved in the process—the problem was, no one knew what or where those chemicals might be.

This argument could only be resolved by closely monitoring the process of digestion. Which requires watching what happens to food as it passes through the entire the digestive tract. That's difficult to do without cutting holes in people, and darn near impossible to do without killing them. It was there that the search for truth stalled until 1822, when Dr. William Beaumont at last found the answer—bleeding out on the floor of a trading post.

DR. BEAUMONT . . .

Born in Connecticut in 1785, Beaumont didn't receive what we would consider a traditional medical education . . . or much of one at all. That's not to say that he was a quack or a fraud by any means. At that time, it was common for many doctors, Beaumont included, to begin their medical careers by "reading under" for a mentoring physician for a couple of years. In layman's terms, that means doing a lot of book reports. In 1811, Beaumont began a year-long apprenticeship that ended with one heck of a final exam: the War of 1812. The war brought Beaumont plenty of on-the-job training as a surgeon's mate and, later, as an army surgeon.

By 1822, he was stationed at Mackinac Island, at the northern end of Lake Huron. Originally built by the British during the Revolutionary War, Mackinac was the site of a couple of battles in the War of 1812. A decade later, however, it was a more peaceful hub of the fur trade. More importantly: It was isolated.

. . . MEET MR. ST. MARTIN . . .

Alexis St. Martin was a French Canadian who roamed the island, collecting furs and selling them for the American Fur Company. Fur traders were seen as rough and wild men who spent every moment they could gambling, drinking, and fighting. This was more often true than not, and St. Martin was no exception. We can assume that, under normal circumstances, he and Dr. Beaumont would never have crossed paths—let alone strike up a conversation with each other. Unfortunately for Alexis St. Martin, however, his life circumstances were about to become incredibly and uncomfortably abnormal.

. . . AND HIS GAPING ABDOMINAL WOUND

In June of 1822, some customers were goofing around with a loaded shotgun when, just like that firearms safety officer warned you about, it went off unexpectedly. Alexis St. Martin just happened to be in the wrong place at the wrong time, just three feet away from the muzzle. A load of buckshot tore into his abdomen, taking out two ribs, a piece of his left lung, and most importantly for our story, leaving a gaping hole in his stomach from which his breakfast immediately spilled.

Beaumont arrived at the grisly scene and instantly knew that his new patient had slim chances for survival. Moreover, Beaumont had no hope of calling in backup; he was the only doctor on Mackinac Island (remember: isolated). He did the best he could for St. Martin, removing pieces of shattered rib, squishing his lung back into place, cleaning out the wound, and applying a poultice to aid with healing.

Then, the two men waited.

Over the next six months, St. Martin fought to stay alive with Beaumont by his side. The wound (inevitably) became infected, and Beaumont did his best to clean it out and try newer and smellier poultices on it. After that, St. Martin developed pneumonia, and Beaumont fought back with one of medicine's oldest and most consistently ineffective cures: bloodletting.

At one point, out of desperation, Beaumont even administered an emetic (something that makes you hurl) in an effort to balance the patient's humors. The vomiting did not go exactly as planned. . . . It came out of the big hole in his stomach. You probably guessed that, but we didn't want to risk anyone being spared that particular mental image.

Beaumont tried multiple times to sew up the wound, but it was hard to keep closed and even harder to keep clean. Beaumont was stuck in a cycle of opening, infection, and reclosure until St. Martin finally decided that he had had enough. The hole was there to stay.

THE SOMEWHAT DISGUSTING ROAD TO RECOVERY

Fast forward another six months or so. With the help of a rectally-administered nutrition regimen, St. Martin was finally on the mend. One little problem remained, though—the hole in his stomach was still open. He eventually was able to swallow food again, but unless he kept a hand over his abdomen, he (to say nothing of anyone near him) was in for quite a mess at mealtimes.

While there was indeed a hole that connected the inside of his stomach with the outside, the walls of the tract had sealed. It's a phenomenon known today as a chronic gastric fistula. Imagine his stomach as New York City and the outside world as New Jersey. Now imagine that the Lincoln Tunnel runs through his abdominal wall. That's a fistula.

Despite this unexpected piece of abdominal infrastructure, St. Martin had made a good recovery to the point that he no longer required the good doctor's constant medical attention. The problem was that he couldn't just return to his old life. Paddling around lakes and skinning carcasses was taxing work in the best of health. Add an expressway to the digestive tract, and it's a lot closer to "suicidal."

Where St. Martin saw a dilemma, Beaumont saw an opportunity. He offered the fur trader a job as his handyman, working and living on his property and doing whatever odd jobs the doctor needed. The oddest of those jobs began with the doctor's request that St. Martin allow him to experiment on his gastric fistula in order to finally unlock the mystery of human digestion.

Alexis was reluctant at first, fearing a relapse

I have heard of many inappropriate relationships between employer and employee, but this . . . this is really up there, isn't it?

"Oh, that? That's Alexis, he just . . . helps out around the place. You know, he does a bit of cleaning, helps me keep my tax documentation in order. That's it . . . well, from time to time, I stick my hand wrist-deep in his tummy, and rummage around his organs like an elderly widow looking for a flashlight in her junk drawer during a blackout, but other than that, it's all business."

from which he might not recover. After a year of persuasion, though, he finally gave in to Beaumont's pleas.

THE TUMMY TRIALS

Beaumont began with pretty straightforward experiments: He would take a bite-sized chunk of food, tie a string around it, and then insert the food into the hole in St. Martin's side. He'd leave it there for some period of time, then remove it and examine it. He meticulously noted how different foods and varying lengths of time in the stomach affected the results.

Thanks to Beaumont's scientific rigor, we know the first round—or "course," if you will—consisted of raw salted lean beef, raw salted fat pork, fresh beef, corned beef, stale bread, and raw cabbage.

In addition to the food trials, Beaumont took samples of gastric juices, and measured the stomach's temperature at different times of day and points in the digestive process, in order to try to reproduce stomach conditions externally. We're not able to ask St. Martin personally, but he likely would have rated the sensation between "very, very unpleasant" and "[barf noise]."

The trapper was able to put up with all this for about a month before Canada started looking pretty good to him again, and he returned home.

Not the sort of man to let an incurable gaping wound define him, St. Martin found odd work here and there, started a family, and sat down to dinner every night with his wife and kids around him and his hand over his middle.

Beaumont, on the other hand, had apparently not gotten the same degree of closure on his end of the relationship.

I WANT YOU BACK

While tending to his duties as a surgeon and an officer, he continued to write letters to St. Martin, imploring him to return so that they could resume the experiments. It wasn't just a chance to advance medical science that preoccupied Beaumont. That was nice and all, but St. Martin also represented a shot at legitimacy for the hastily-trained surgeon. This was the case that would secure his legacy, and buoy his family's finances in the years to come.

He could not let this opportunity pass by.

WELCOME HOME

With the promise of money to support his wife and children, St. Martin reluctantly agreed to return to the doctor's care, and submit to further scientific inquiry. This time around, Beaumont added field trips for St. Martin to his scientific method, so as to observe how changes in temperature and weather affected his gastric juices. But it wasn't all sunshine and fresh air—he also started forcing the already nauseated Alexis to exercise so that Beaumont could observe how that might change the composition of the juices or the speed at which food broke down. St. Martin's irate response to the exercise plan (P90Blech? CrossPuke?) inspired the doctor to start a new series of investigations into how changes in mood impact digestion.

"Well, we gave it a go, but I'm just gonna live with this stomach hole. I'm just gonna ride this thing out."

Maybe he just thought it could be a quirky affectation. You know "Oh, you have to meet my friend Alexis; he's got the most ironic stomach hole." It's not that odd. I know a lot of people who are into extreme body modification. It's just like saying "I decided I want to stretch my earlobes way out; that's just my choice." Except St. Martin's saying "I wanted a hole down there in my tum-tum, that's just my choice—and oh my god, is that what chewed-up fried chicken looks like?"

Through it all, Beaumont diligently recorded all of his findings. Every bite of food, drop of alcohol, daily weather condition, and internal gastric temperature was dutifully copied into his notebook. Which gives the modern reader an appreciation for his scientific rigor, as well as the opportunity to critique someone else's daily food choices. For example, nothing makes you feel better about a pizza bender than reading that on October 24th, 1833, Alexis St. Martin ate "a pint of soft custard and nothing else."

Despite the somewhat intimate nature of their association, the personal relationship between the two men only got stormier. One of the reasons Beaumont's study took so long to complete was that it was interrupted by frequent arguments that resulted in periods of estrangement. Beaumont paid for the whole St. Martin family to come stay with him at Fort Crawford in 1829, and things went well for awhile. However, in 1831, a particularly nasty squabble lead to a year's separation.

REUNITED AND IT FEELS SO GOOD

When the two reunited in 1832, Alexis left his family behind. For several months, the two men stayed in a hotel in Washington, DC, where Beaumont continued his experiments with different foods as well as different quantities of some items—such as placing twelve raw oysters into St. Martin's stomach at the same time.

Perhaps that was the last straw, because St. Martin eventually decided that their D.C. rendezvous would be their last. The men returned to their respective families and went about the rest of their lives. For many years, Beaumont continued to write St. Martin hoping to beg, bribe, or cajole him into returning for more experiments—all to no avail.

Beaumont published his findings in 1833, proving conclusively that digestion had something to do with gastric juices based on those he had collected from St. Martin's stomach over the years. He also continued his career as a physician, now of greater renown than before, practicing until his death in 1853. He lives on though, as the "Father of Gastric Physiology."

For his part, St. Martin never saw Beaumont again. He joined up with a traveling medicine show for a while, being displayed to awed and disgusted crowds as a human oddity. After that stopped being fun, he was able to return to fur

JUSTIN Wait, breaking up and getting back together? Confined spaces for lengthy periods? Extremely intimate physical contact? I've seen enough romantic comedies to fill in the blank spaces here: These two were in love.

SYDNEE No, I'm not saying that at all. They were both of completely different social status; they didn't even speak the same language.

JUSTIN Wait, I'm confused, I thought you were laying out the reasons that they wouldn't like each other. This sounds more like the trailer for a Katherine Heigl movie.

SYDNEE No, not like opposites attract. I mean they literally spoke different languages.

JUSTIN Oh, okay, my mistake. So, you're actually 100-percent describing the Colin Firth part in *Love, Actually*.

trading, and eventually became a farmer. He lived to the ripe old age of eighty-three, fistula and all.

Despite St. Martin's refusal to respond to Beaumont's invitations those last twenty years before his death, there is a more satisfying conclusion to their relationship: In 1856, while touring the country with a medicine show, he made a point of visiting St. Louis, Missiouri, to pay respects to Beaumont's widow and son, Israel.

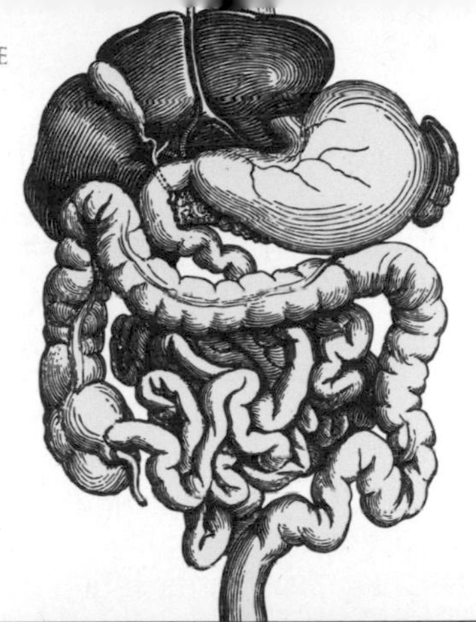

JUSTIN VS SYDNEE

JUSTIN I think I can take story from here, Syd. St. Martin turned to Beaumont's son with that look like, "What do you think? For old time's sake?" and the kid looks up at him like, "You know I'm down," and he just buries his little fist in St. Martin's gut. Then they jump in the air and do a freeze-frame high five. (The kid's hand is still in his gut, natch.)

THE END . . .

SYDNEE Well . . . no, but a friendship did blossom between Alexis and Israel from that initial visit, and the two would continue to exchange letters for the rest of St. Martin's life, but—

JUSTIN Sydnee, I said freeze-frame high five. You can't put anything after a freeze-frame high five. That's the law.

SYDNEE But—

JUSTIN THE LAW.

So, What Have We Learned?

While Beaumont's work helped to shed some light on the process of human digestion, it wasn't necessarily a massive breakthrough. A great anecdote for doctors eager to get themselves shunned by polite company? You bet. The moving story of intimate platonic love between a man and his friend's stomach wound? You betcha. But not earth-shattering.

Luckily, we no longer have to hope for extremely fortuitous gunshot wounds to study the gut. We can take a peek inside the stomach or intestines using a very small camera in a procedure known as an endoscopy.

We also developed surgical procedures that allow us to fix a fistula. It works . . . basically how'd you'd imagine sewing up a big hole in someone's stomach would work. If the fateful shotgun had happened today, St. Martin would likely have been able to keep fur trading, and Beaumont could have moved on with his life. If that idea bums you out, let us cheer you with the fun fact that the test we use to find a fistula is called a "fistulagram," which sounds like the absolute worst way to wish your grandparents happy anniversary.

For the vast majority of human history, mental illness was seen as an utterly hopeless diagnosis. By the late 19th century, we began to develop an extremely basic understanding of how the brain functions. Though it was a considerable leap forward for medicine, it also lead to an equally massive misstep: the lobotomy.

The Innovators

The first lobotomy was performed in the late 1880s by a Swiss physician named Gottlieb Burckhardt, who presided over an asylum. He theorized that he could calm patients (not fix them, mind you) by removing certain parts of their brains. Of the six patients he worked on, one died a few days later (which is kind of the ultimate calm, really), but the others did become more docile. Burckhardt thought the work showed promise, but lobotomies didn't gain much traction until the mid-1930s, thanks to Portuguese neurophysician António Egas Moniz.

Moniz had severe gout, which impacted the use of his hands, but he didn't let that keep him from his dream of cutting people's brains up. He believed that mental illness was caused by faulty brain connections, and he enlisted the help of a surgeon named Pedro Almeida Lima to help him prove it with a procedure they termed a "leucotomy," severing the brain's connective white matter.

The Treatment

Moniz and Lima's first test group consisted of twenty patients who presented with anxiety, schizophrenia, and/or depression. Lima initially injected ethanol into the frontal lobe which . . . well, to put it technically, it messed the lobe up real good. They later developed a wire loop to sever the white-matter fibers, which was not only more precise, but way better nightmare fuel for early 21st-century readers, conveniently for us.

Of the first twenty patients operated on, Moniz claimed seven were cured, seven were somewhat improved, and six were unchanged.

That last bit is how you know they were making it up as they went along. Seriously, Lima? Six were unchanged? It's never happened to us, but we can pretty much guarantee that someone chopping your thinkmeat into bite-sized pieces is going to do something. Maybe you just start mixing up blue and yellow or thinking Carrot Top is hysterical, but you're not getting out of it unscathed.

This leucotomy was briefly in fashion, hailed as a miracle cure. In fact, Moniz won the 1949 Nobel Prize for Medicine for inventing the procedure.

How'd that work out?

Uhh, very badly! American neurologists Walter Freeman and James Winston Watts refined the procedure, christening it the "lobotomy," and making it even more horrifying. In Freeman's innovative twist, as it were, the doctor hammered a metal instrument not unlike an icepick into the socket above the patient's eyeball, and then wiggled it around to sever connections in the frontal lobe of the brain. Though designed to relieve paranoia and anxiety, it left many patients in a vegetative state. Still, that didn't stop Freeman from pushing lobotomies . . . hard.

Freeman himself lobotomized 2,900 people, including nineteen children. He would perform lobotomies for the media, sometimes inserting two picks at the same time to show off his ambidexterity. During one such demonstration, a patient died on the table, but Freeman barely even paused before moving onto the next patient.

If that's not disgusting enough, it's worth mentioning that 40 percent of Freeman's surgeries were meant to "fix" the sexual orientation of perfectly healthy people who just happened to be gay.

Eventually, mercifully, public outcry against the effects of lobotomies knocked them out of favor.

NOT-SO-FUN FACT: Before hanging up his ice pick, Walter Freeman even lobotomized Rosemary Kennedy, JFK's sister in 1941, when she was just twenty-three years old. By the time Freeman was done with her, she had to be institutionalized, as she was incontinent and unable to speak or walk. The debilitating issue this horrifically risky procedure was intended to fix? Mood swings and learning difficulties.

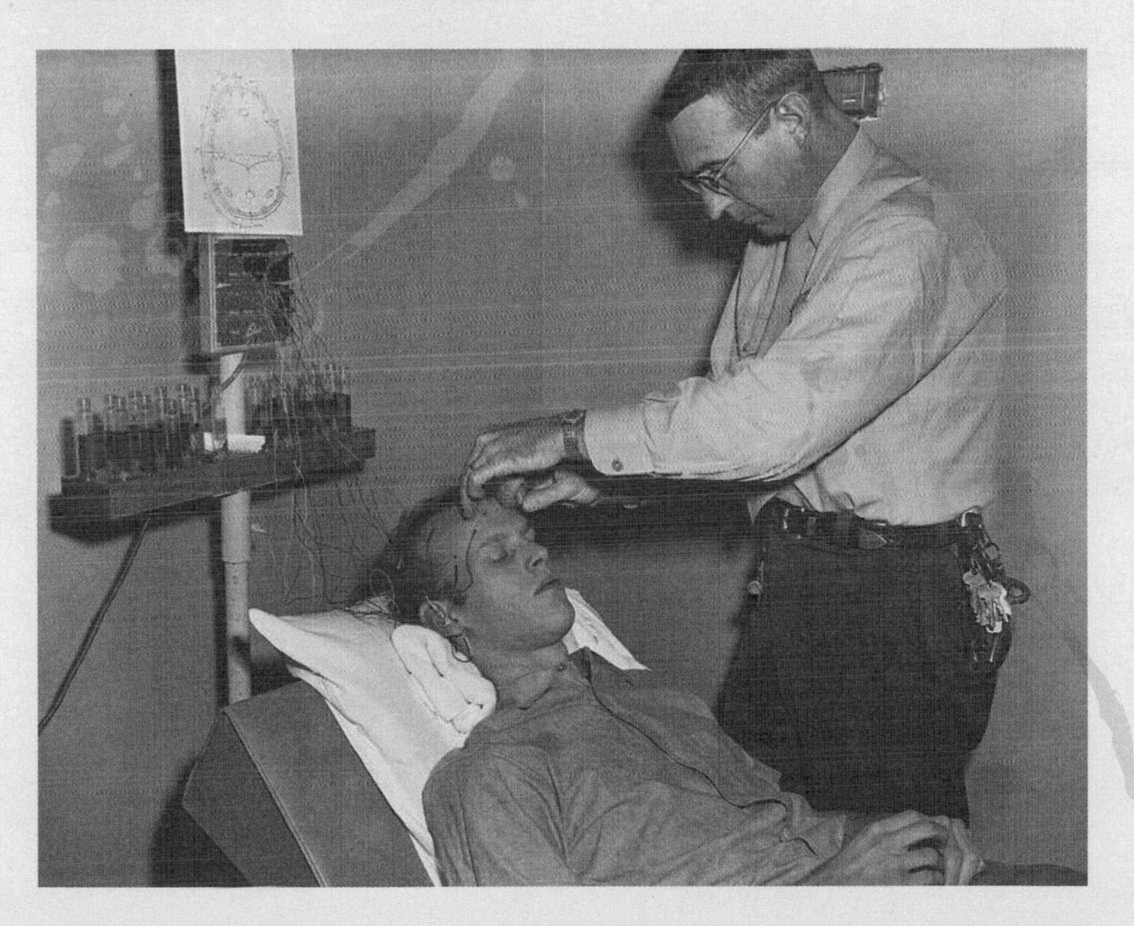

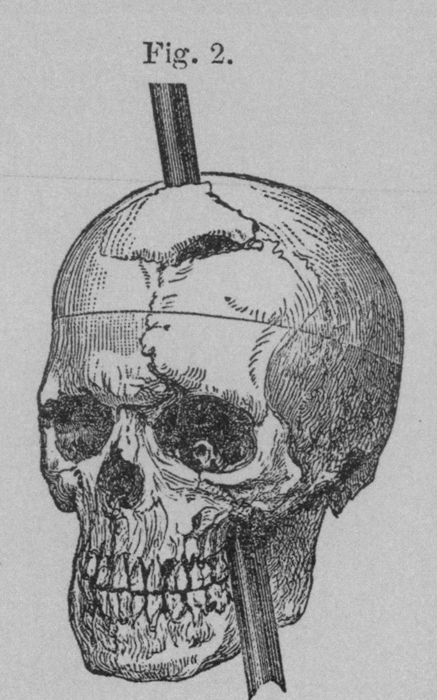

The Unkillable Phineas Gage

Gather 'round, children. Ol' Pappy Sawbones has a tale to tell. Hezekiah, put down Pappy's relaxing tonic, that's not for young 'uns.

Kids, let's take a break from all this history and let me bend your ear with the story of a man named Phineas Gage. He was born in 1823, the eldest of five children of Jesse and Hannah Gage of Grafton County, NH. And while his tale may sound tall, I will assure you that it's all true—sure as my name is Ezekiel Stilton Sawbones.

Prior to the incident that would make him famous, Gage was a healthy young man, real strapping, like your cousin Hunkus. We even have evidence from his family doc, Dr. Harlow, who wrote that he was "a perfectly healthy, strong and active young man, twenty-five years of age, nervo-bilious temperament, five feet six inches in height, average weight one hundred and fifty pounds, possessing an iron will as well as an iron frame; muscular system unusually well developed—having had scarcely a day's illness from his childhood to the date of [his] injury."

He was not a well-educated man, and instead used his strength to further his trade, and there's no shame in that, let me tell you. He was working in railroad construction on the Hudson River Railroad and eventually got to be a blasting foreman.

ONE LITTLE SPARK

On September 13 of 1838, Phineas was hard at work helping to build the Rutland and Burlington RR in Cavendish, Vermont. He had made a hole in a rock, filled it with gunpowder and a fuse, but hadn't gotten to the sand yet. He was tamping it down anyway, which may or may not have been okay to do without the sand, I'm not clear on that point.

Regardless, he was using a 3-foot-7-inch long, 1.25-inch-diameter iron rod that weighed 13.25 pounds to do this. It probably seems like your Ol' Pappy is going into too great a detail as to the specific dimensions of this rod but . . . uhh, it gets important real quick.

(On a side note, Gage made his own custom tamping rods at some point, but this was not one of

Now, kids, as I'm gonna reckon you've never done any work as a blasting foreman, let's talk a second about this job:

So, in order to construct a railroad, you have to blow things up that are in the way of said railroad. This was done by boring a hole in a rock, putting in gunpowder and a fuse, pouring sand down into the hole, and then tamping it down with a big iron rod, called, you guessed it, a "tamping rod." Once everything is packed nice and tight, you light the fuse and run. You run like a ranch hand chasing a dairy maid that stole his lucky penny, and let me tell you: That's fast. Also, you probably yelled "Fire in the hole!" or something like that.

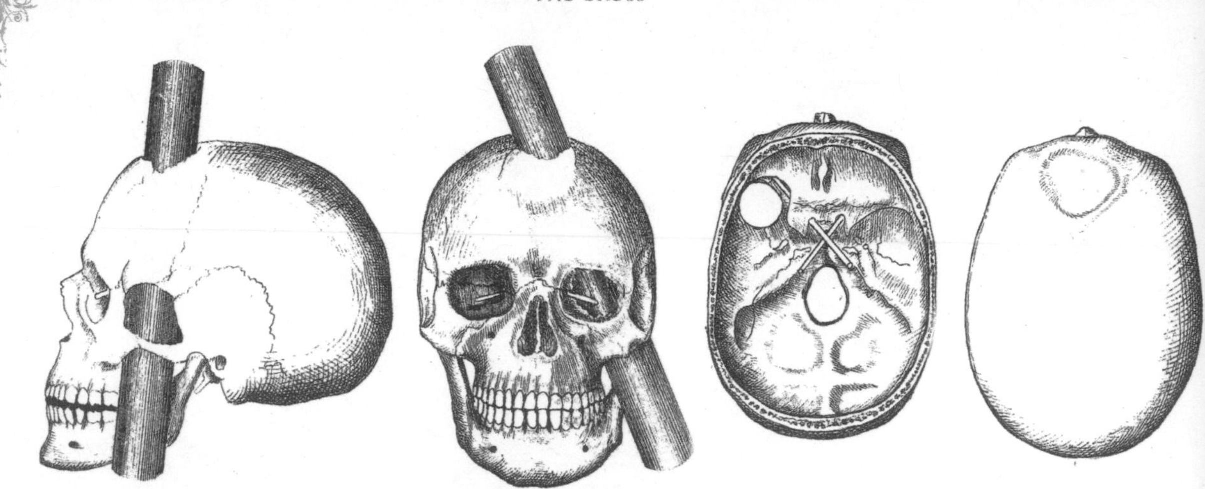

them. This one was made for somebody else who liked it tapered a bit at one end. He had borrowed it. That tapered end was real lucky for Gage, but look at me getting ahead of myself again.)

Unfortunately, Gage's loaner rod struck the rock as he was tamping, and made a spark. Now, if you're trying to win immunity on *Survivor*, that's real good news. But Gage's spark didn't land on a pile of coconut fibers that Jeff Probst gave him. It landed on gunpowder—and then it exploded.

The rod shot up like a rocket, clean through the head of poor Mr. Gage. It entered on the left side below his eye and passed up behind it to exit through the top of his head. It was found about eighty feet away, covered in blood and brains.

And then he died.

HE DIDN'T DIE

No, no, kids, come on back. Pappy was just funning. That wouldn't be much of a story, would it? I mean outside of its value as a cautionary tale about being real careful with your tamping rod.

After that mean ol' tamping rod shot through his head, Phineas Gage fell down, convulsed a bit, and then sat back up. He was even talking to the crowd of people who had gathered around, who, let me tell you, couldn't have been more surprised. Someone gave him a ride back to the hotel where he was staying, and believe it or not, he was alert and awake the whole time.

Finally, a doctor was called for, and when Dr. Edward Williams arrived, he found Gage sitting on the porch. The following is from his account of the case, made at the time:

"When I drove up he said, 'Doctor, here is business enough for you.' I first noticed the wound upon the head before I alighted from my carriage, the pulsations of the brain being very distinct. The top of the head appeared somewhat like an inverted funnel."

Now this doctor was a man of science, so he didn't necessarily believe Gage at first, but Phineas decided to play his trump card:

"Mr. G. got up and vomited; the effort of vomiting pressed out about half a teacupful of the brain, which fell upon the floor."

Now that is one heck of a way to win an argument, ain't it?

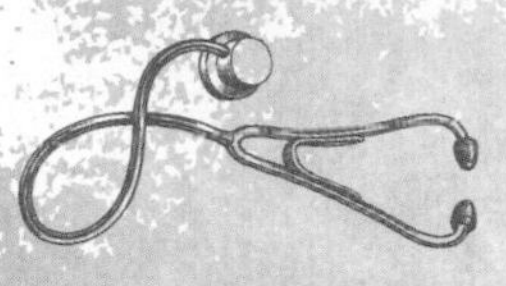

Sydnee's Fun Medical Facts

We're much better at fighting infection today, but a fungal abscess of the brain is a serious condition, no matter what century you're in. The walled-off pocket of fungal material can be difficult to penetrate even with intravenous medications. And the drugs that will work can be fairly toxic with their own suite of side effects. One of the treatments we use most commonly is amphotericin B, but in medical circles, it's frequently referred to as "amphoterrible." That should give you a pretty good idea of how much fun it is.

Doctor Williams called for Gage's own physician, Dr. Harlow, to come help out. Harlow would later describe it all as "truly terrific," and isn't that just a doctor for you?

So Williams and Harlow—who, by the way, were really just making this up as they went—shaved his head, removed the clots and bone bits and an ounce or so of brain, and then bandaged the whole thing up loosely, as well as his face, and all the burns he had gotten on his arms.

Remarkably, Phineas still seemed okay. He didn't really want to see his friends just yet, but he said he'd be back at work in a few days.

> Okay, I'm sorry to put another sidebar right after Sydnee's. Our layout guy, Scott, probably feels like he's got an intracranial fungal infection of his own, but with the inclusion of "cool slang terms for brain fungus drugs" has my wife officially stretched the definition of "fun fact" to its breaking point? I submit that she has.

THE LONG, LONG ROAD TO RECOVERY

That didn't happen, of course. In fact, Gage had a long and difficult recovery ahead of him. Keep in mind, this was back before we had all these fancy antiseptic techniques and antibiotics. Heck, we didn't even wash our hands all that often. So it's no surprise that Gage developed fungal infection at the site of the injury. The swelling and drainage and general inflammation, alongside the actual fungal spores that were taking up space, resulted in stuff bulging out of the head wound and his eyeball poking out of his face a bit. It was real cute.

For a medical man of that era, Dr. Harlow seems to have been pretty good with all this stuff, such as keeping things open and draining when they should be and even cutting open an abscess that formed on Gage's face and impeded recovery. These techniques were quite advanced for the time. If you were to make a list of reasons Gage survived his brush with death, Dr. Harlow and old-fashioned divine providence would be in a shoving match for the top spot.

By November, Phineas had recovered enough to return home, and even work around the farm for a half-day at a time. His return to the simple life was short-lived, however, as others in the medical world began to take notice of him. He was invited to Harvard so the doctors there could poke and prod him and, just maybe, figure out for themselves if his tall tale was true.

HOW PHINEAS GAGE WENT TO HARVARD

This was probably one of the earliest examples of a patient being brought into a teaching hospital strictly for the educational opportunity and not for any additional treatment or care. He also pursued some less educational, but perhaps more lucrative, offers to appear in Barnum's American Museum, as well as multiple paid public appearances throughout the United States.

And then, as so often happens, the public's interest in that amazing man with the hole in his head began to wane, and poor Phineas had to find alternative means to support himself. He worked for a bit at a stable in New Hampshire, and then as a stagecoach driver in Chile. He never loaned himself out as an exotic cupholder, which is a credit to his dignity, I think.

In 1860, poor Phineas began having seizures. Unable to work, he returned to San Francisco where he had family, and died of complications related to the seizures in May of that year, twelve years after the rod went through his head. Harvard Medical School now has the skull and iron on display for the public to view in its Warren Anatomical Museum.

It's already a heck of a story, but Phineas Gage provided history with more than just a sideshow: He helped us understand how the brain works.

WILL THE REAL MR. GAGE PLEASE STAND UP?

This contribution all goes back to Gage's behavior after his accident, which some people said turned him a little squirrely. By all accounts, prior to that tamping rod taking a load off his mind, Phineas was a nice guy, a hard worker, well-liked by both bosses and buddies, and generally considered a good egg. However, there are reasons to believe this may have changed as a direct result of his unfortunate accident. Contemporary reports tell us that he became vulgar and inappropriately sexual, profane, violent, impulsive, and generally a jerk. Heck, even by his own doctor's estimation he was "no longer Gage" after the accident. He wasn't even hired back by the railroad, despite his excellent employment record.

To be fair, there are inconsistencies in this part of the story. Some people wrote that he was still as nice a guy as ever, just with a slight depression in his skull and a bit of a droopy eyelid. Supporting this argument is the fact that he managed to land a job as a stagecoach driver, which would have demanded hard work, organization, and drive. In addition, he was in another country and had a language barrier to overcome. It seems unlikely that he could have managed all this if he was completely unhinged all the time. So, the question remains as to how much Phineas Gage really changed after a significant part of his frontal lobe was blown away.

Why is it important? Well, a lot of our early understanding of the brain was based on this story. The frontal lobe is considered the seat of our personality, and this story is frequently cited as evidence. In fact, you'll find this tale I've been spinning for you in two-thirds of all psychology textbooks.

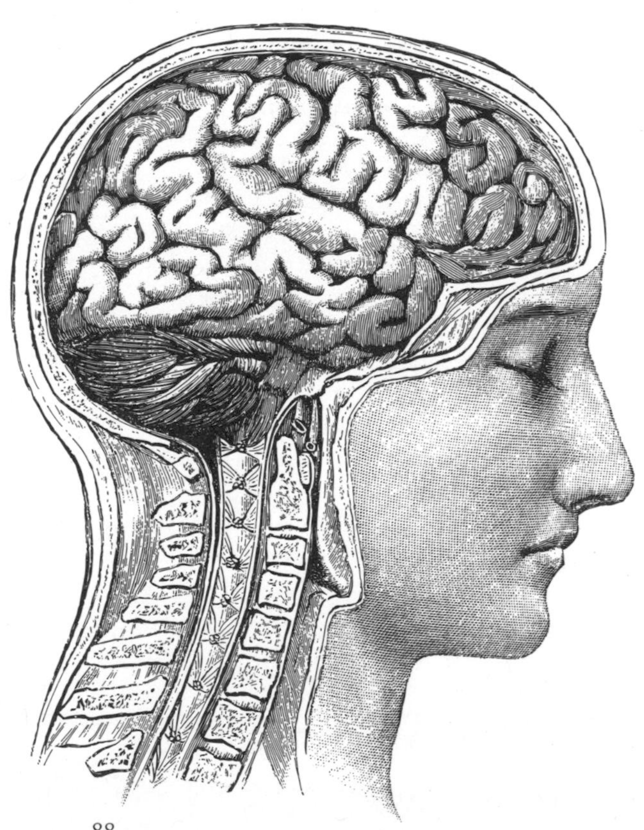

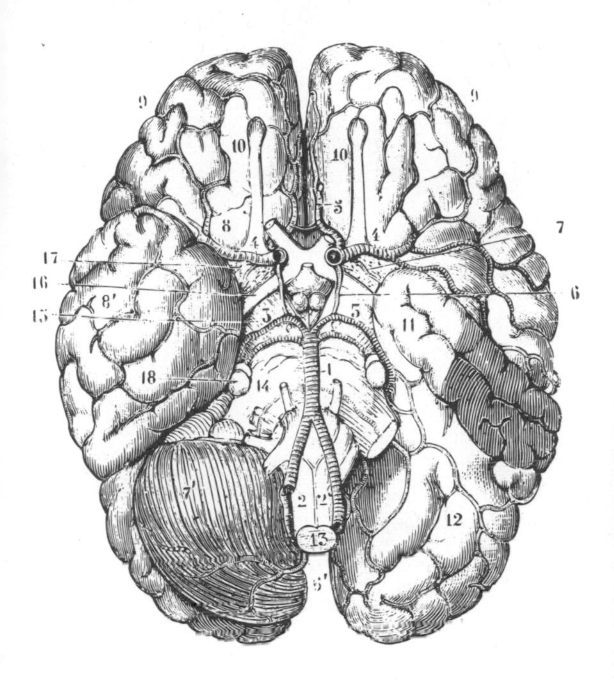

AN INSPIRATION TO US ALL

This same story would also inspire plenty of budding surgeons. This was seen as proof that it was possible to perform surgery on the brain without killing the patient, and surgeons were itching to prove rise to the standard set by Phineas' rogue tamping iron.

Of course, it wasn't all pearls and roses. Gage's story was also used to try and prove fake theories such as phrenology. Many who believed in the practice of predicting personality and behavior based on head lumps interpreted the results of this accident as evidence that the iron had destroyed his organs of benevolence, veneration, and compassion on its way through. If that just sounds like a bunch of hooey to you kids, well . . . it's only because phrenology is completely made up.

Now, you kids get on to bed; Pappy's stories are coming on, and I'm aiming to give my feet a good soaking. Before you head upstairs, I want to leave your noggins with something to chew on: The tall tale of Phineas Gage isn't just a wild way to spook your weak-stomached friends. I mean, it is that, too—after all, the man caught a rod through his brain. It's something else. But Gage also gave us a deeper and more complex understanding of the human brain, and inspired a generation of researchers and physicians. Pretty darn impressive for a guy who just ducked too slow in *just* the right way.

SO, WHAT HAVE WE LEARNED?

The tale of Phineas Gage is one of the earliest records of someone surviving a traumatic brain injury . . . and when we say the word "traumatic," boy, do we mean it. (He was extremely fortunate to survive the accidental lobotomy. So are we—or we'd not have this chapter to write!)

Gage's injury is also a good example of how brain damage, especially to the frontal lobes, can lead to changes in behavior and personality, though some unconfirmed accounts suggest his behavior improved in the years before his demise. Neuroscience has borne out many details of Gage's injury; in 1865, Pierre Paul Broca discovered that, for right-handed people, a section of the the left frontal lobe of the brain (the inferior frontal gyrus, or Broca's Region) is part of processing speech. Further research in the 1860s, carried out by David Ferrier and John Hughlings-Jackson, added evidence to localization of functions in the brain.

All this supports the idea of the damage to Gage's brain being responsible for his changes in mood and speech. No matter what, though, we're also pretty sure that having an iron rod blast through your brain will leave anyone feeling a bit testy!

So What's the Deal With: PHRENOLOGY?

Your bumpy head doesn't tell us that much about you. Sorry!

When Did This Become a Thing?

Like a lot of medical and pseudo-medical theories we've discussed, the thinking behind phrenology can be traced back to Hippocrates. The general consensus at the time was that the heart controlled everything in your body. Hippocrates put forth the revolutionary, and shockingly correct, theory that the brain was actually running the show. Galen agreed with this, but then broke old-timey medicine's triumphant one-win streak by casually adding that the brain is actually a cold, wet ball of sperm. Nice try there, Galen.

One thing Galen did get right was that certain regions of the brain were linked with individual parts of the body. Its relationship to personality, however, was more of an enigma. There wasn't a lot to go on in the way of answers until 1796 when a German physician named Frans Josef Galle invented "cranioscopy," the study of the skull and how it correlated with personality types. (Actually, this didn't provide any answers either. But for a little while, people thought that it did.)

Gall's collaborator, one Johann Spurzheim, came up with the term "phrenology," meaning "the study of thought." The name stuck. It was still very bad and wrong, but hey, at least it was a little bit more marketable.

Just as Galen had connected specific spinal nerves to their corresponding muscles, Gall and Spurzheim were certain that different parts of the brain controlled different aspects of personality and behavior. The problem was that they had no way of examining the brains of living subjects to help them figure this out without making them, you know, *not* living anymore. That's where phrenology came in.

How's That Work?

The brain isn't divided as cleanly by function as Gall imagined, but his theory was still pretty solid. (See if you can detect the exact moment where it gets decidedly less so.) Let's say that you're a very kind person. Gall theorized that the portion of your brain associated with kindness would be bigger than in a less kind person.

Except Gall also believed that each skull fits the brain underneath it perfectly, like a glove. Based on the fact that a baby's skull is somewhat malleable, he guessed that if as the "kindness" sector of an infant's brain got larger, their skull would adjust to accommodate it. (Like how the Grinch's heart broke the X-ray machine when it grew three sizes that day?) So, by measuring the skull, and feeling bumps or other variations, traits could be described and predicted. He diagrammed a brain made up of twenty-seven "organs," each with a behavioral or personality trait, including the self-esteem organ, the wit organ, the love-for-our-children organ, and the conjugal-love organ (in the brain. Not . . . you know, the other one).

We have reached the exact moment, by the way . . . no, we're past it.

Most of Gall's early subjects were criminals, and in studying them, he was specifically looking for either the murder organ or the theft organ. His method was less than scientific. He would find the bump he associated with the villainous impulse on one cranium, and then just look for similar lumps on others. While this may not seem terribly unreasonable, he would then only count the subjects who fit his theory and discard all of the rest, thereby using bad science to prove his bad science.

At the peak of phrenology's popularity, you could pay to have a prospective spouse or employee's scalp fondled to be sure they didn't, say, have an overgrown murder organ. While most treated phrenology more as a frivolity like fortune-telling or palm reading, there were those who believed it had practical applications. Some who studied criminality used phrenology to justify rehabilitation over punishment. You can't shrink your large murder organ, but maybe you could channel those urges into being a butcher.

In 1905, a guy in Wisconsin, Henry Lavarie, streamlined phrenology with the psychograph, a helmet that was supposed to electronically read your personality in thirty-two different organs. You got a printout rating you on these various factors as well as suggested careers. If you see "butcher," be very afraid.

So, What About Today?

Phrenology was largely discredited during the 20th century, and just in time, as Europeans had begun using it as a way to "scientifically" justify racism, which is both totally gross and wholly unsurprising.

In the modern era, if your doctor is examining your head for lumps, you've likely had some sort of trauma and they're checking it for swelling . . . that, or they're actually a criminally misinformed, racist, pseudoscientist weirdo.

If that's the case, we're sorry you had to find out this way. Take heart; there are plenty of other doctor fish in the medical sea.

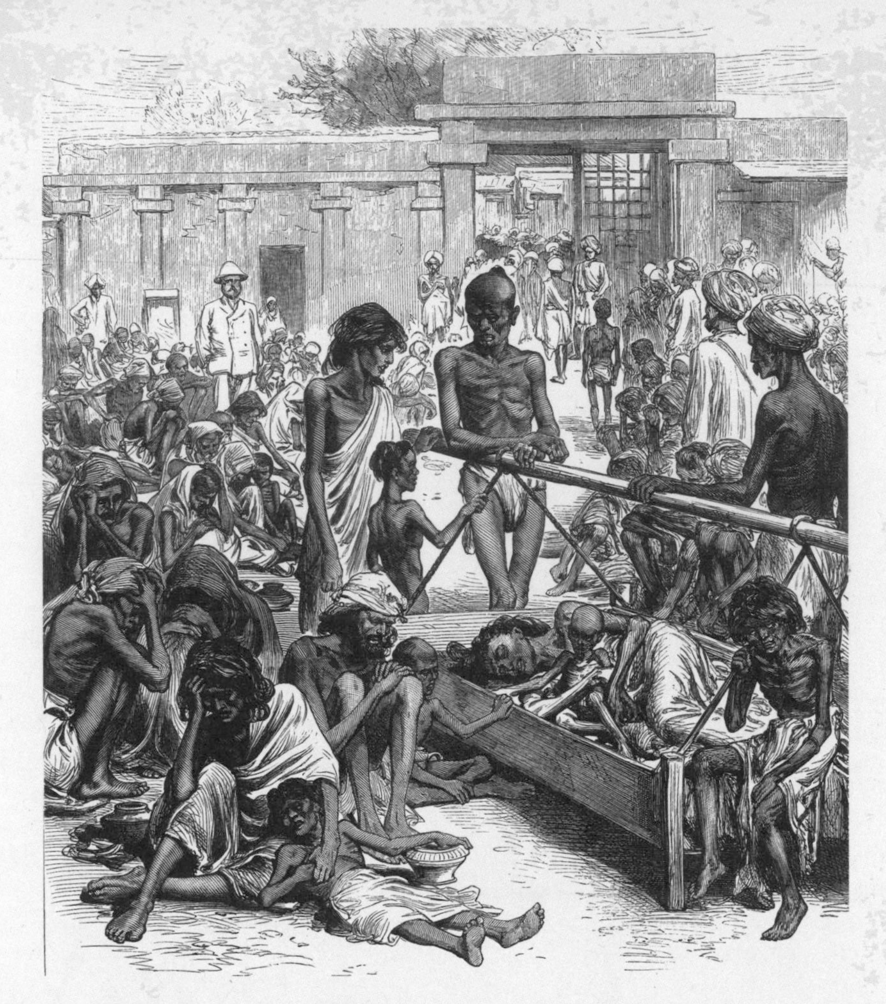

THE MAN WHO DRANK POOP

"Oh yeah, cholera! I know cholera!
That's the one you caught
on the Oregon Trail!"

Be thankful that the closest you've likely ever come to cholera exposure was watching your own ill-fated westward trek through the monochrome screen of an Apple II. In the real world, cholera is . . . very bad news.

If you want to get crass about it (and we always do), cholera is a vicious bacterial disease that can cause massive diarrhea. At first blush, that might sound like nothing more than a fairly yucky inconvenience, but cholera has been responsible for the deaths of millions throughout history.

HAVE BUG, WILL TRAVEL

The word "cholera" comes from the Greek word for bile, but the disease itself probably started out in India, and has been kicking around since ancient times. Throughout the centuries, this pernicious little bugger has spread influence by traveling via trade routes. Cholera has been responsible for eight epidemics throughout recorded history, the last of which was as recent as 2010, when a 7.0 earthquake decimated the infrastructure of Haiti.

Planning on getting cholera? Well, that's a very bad idea, and frankly, we don't know why you're entertaining it. But you seem to be really into it, so we'll be providing you with a step-by-step guide (see page 94).

Without a handy aid like the one we have so generously provided, historical societies had no idea why the disease was happening, or how to stop it. But they did know enough to stay away when an epidemic started.

In fact, quarantine was the only surefire method of staying healthy in many people's minds, and a number of measures were taken to try and contain the disease. Sometime around the 17th century, for instance, it became maritime law that if cholera broke out on a ship at sea, it was required to fly a yellow flag to indicate the presence of the disease on board. Passengers and crew had to stay aboard in dock for thirty to forty days before they were allowed to come ashore once the yellow flag was flown. During an outbreak in New York City in 1832, 100,000 of the city's 250,000 inhabitants left the city to avoid illness.

If a proportionate number of residents left the city today, we'd be talking about an exodus of 3.4 million people. To give you some perspective, that's . . . a lot of people.

Those who didn't flee the disease tried anything they could to fight it. In the Indian subcontinent, where cholera was endemic, traditional treatments sought to rebalance the humors, and calm intestinal spasms, largely through the use of heavy metals like mercury in the form of calomel—which was both poisonous and had laxative qualities, so it may not have been super helpful. Another remedy, published in an 1864 medical book, called for a tincture of brandy, peppermint, and a hefty dose of our good friend opium. Essential oils and spices were used as well, but this may have been to combat the smell as much as anything else.

Sometimes heat was the only treatment offered, in the form of baths, friction, cauterizing the heel (yeah, your guess is as good as ours on that one), or even using tourniquets on the limbs to cause redness and swelling.

Of course, bleeding was popular throughout history, only supplanted by opium later on.

One treatment account from the 19th century involved putting a paste of lemon juice, iron oxide, and alum over a man's eyes. Here's the account of a Dr. A.L. Cox who witnessed the treatment being administered by an Indian doctor:

> *"The pain it produced vexed and enraged the sick man, and he attempted to strike those around him; the vomitings became more frequent, his attendants fled to avoid his blows; he pursued them; passing by a reservoir of water, which served for the purposes of the garden, he plunged into it and drank with avidity for several moments. They surrounded him, but he remained tranquil in the water. The enormous quantity of liquid he drank, was followed by fainting. He was then removed from the reservoir and put to bed; he slept for eleven hours. When he awoke, the vomitings and dejections had ceased, but he was blind."*

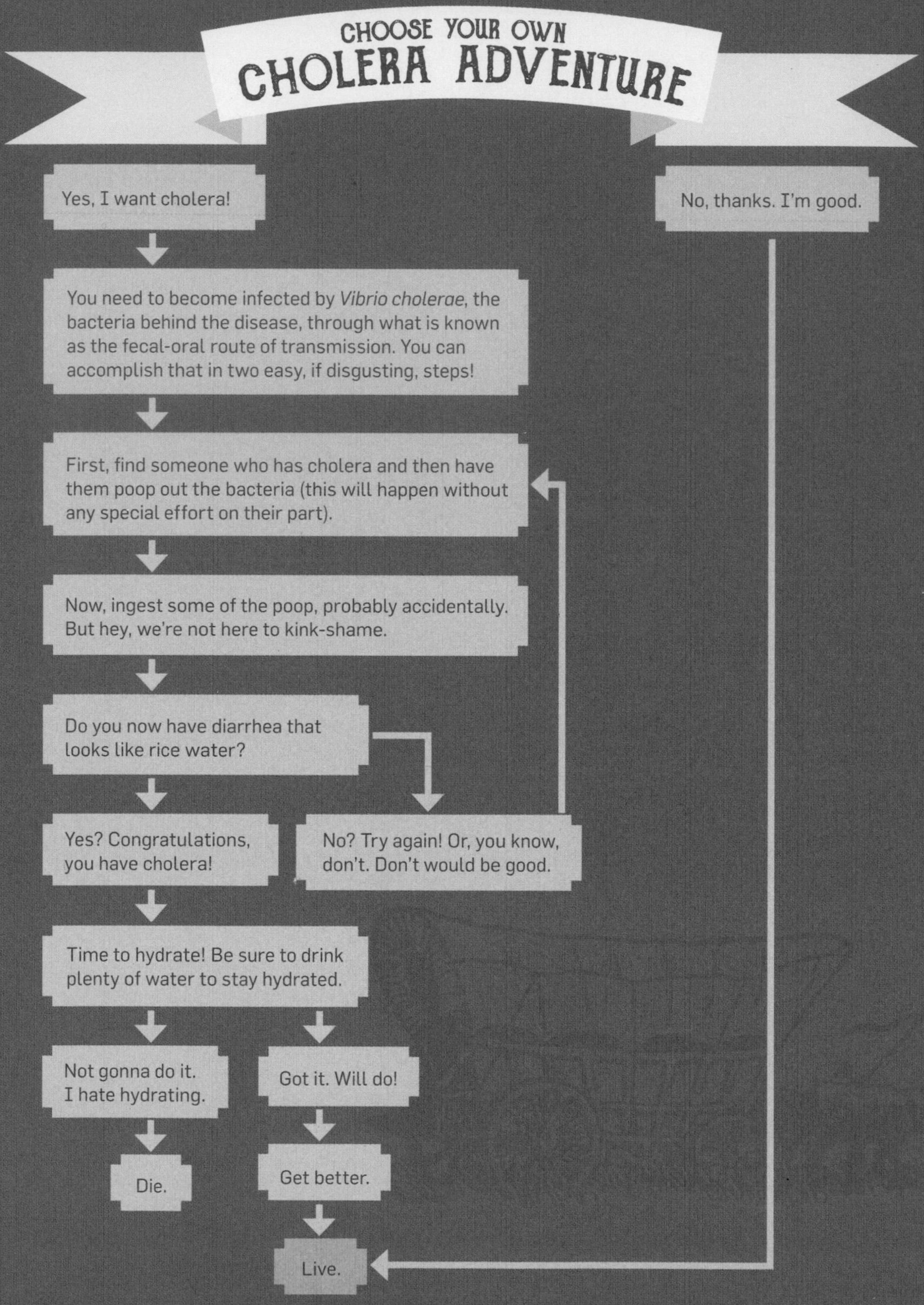

CHOOSE YOUR OWN CHOLERA ADVENTURE

Yes, I want cholera!

You need to become infected by *Vibrio cholerae*, the bacteria behind the disease, through what is known as the fecal-oral route of transmission. You can accomplish that in two easy, if disgusting, steps!

First, find someone who has cholera and then have them poop out the bacteria (this will happen without any special effort on their part).

Now, ingest some of the poop, probably accidentally. But hey, we're not here to kink-shame.

Do you now have diarrhea that looks like rice water?

Yes? Congratulations, you have cholera!

No? Try again! Or, you know, don't. Don't would be good.

Time to hydrate! Be sure to drink plenty of water to stay hydrated.

Not gonna do it. I hate hydrating.

Die.

Got it. Will do!

Get better.

Live.

No, thanks. I'm good.

Well, umm . . . win some, lose some, right? Win some . . . not vomiting, and lose some . . . sight. Seriously, I have to tell you, "win some, lose some" really loses something when you start filling in the blanks, huh?

RED FLANNEL DISEASE PREVENTION

While disease prevention is generally preferable to treatment, that only holds true when the method of prevention is based on sound science. Many people in the 1800s believed that a chilly belly could leave one open to illnesses such as cholera or dysentery. The solution: simply wrap a piece of red flannel around your middle. For many years, soldiers in the British Indian Army were issued two of these so-called "cholera belts" to wear under their uniforms. These persisted as part of standard military issue throughout the 1800s and were even used in some instances during World War I. And in fact, the concept of abdominal chilling as a cause of disease did not completely die out until the 1940s.

Keep in mind, we're talking about how people have gone about treating cholera in the many centuries before antibiotics, intravenous fluids, or oral rehydration solutions. It's still a dangerous disease today, but before the modern era, cholera was absolutely brutal. This unfathomable threat to public health demanded a few brave souls to summon up all of their courage and—

Okay, fine, we're couching. It's just that we're about to tell you about a guy who did something really grody to try to crack the cholera code, and we wanted you to have the full historical context, lest you just think he's, you know, a total grosso freakazoid, rather than a well-meaning scientist.

Historical niceties complete, we'll just say it: This is the tale of the guy who drank poopy water. We'll get to the details, promise. But first, we need to understand why the old dirty bird—sorry, scientist—would have done such a thing.

Design for Cholera Belt

Front

Back

MEDICAL DETECTIVES

Prior to our understanding that germs cause disease, people had a lot of strange theories about illness. We already mentioned that in India, cholera was once thought to be caused by an imbalance of internal forces, or humors. The Europeans and Americans later thought it was more of a miasmatic disease, meaning that it was the result of foul air that drifted your way. They surmised it had something to do with being close to filth and waste. Which wasn't entirely wrong.

Congratulations, old-timey doctors! You accidentally sort of got one half-right! I'm not being sarcastic either; that's pretty darn good, considering just how dumb you've been about basically everything medical! Wait, that wasn't nice either, was it? . . . Wait, wait, come back!

A far less laudable (and far more wrong) idea was that cholera transmission was somehow related to race or cultural background. It was also sometimes blamed on the sufferers engaging in immoral behavior, or perhaps just having the misfortune of being poor. It likely won't come as much of a surprise, then, that outbreaks in the United States would often be unfairly blamed on various immigrant populations.

It was not commonly believed that you could transmit cholera from person to person, because the doctors who were caring for those stricken with the disease rarely became ill themselves. In reality, the doctors probably stayed well simply because they weren't drinking the same contaminated water that their patients were consuming.

In the late 19th century, the growing realization that germs are responsible for at least some illnesses (the not-so-cleverly-named "germ theory of disease") should have changed a lot of these misconceptions. Doctors and scientists were finally understanding that tiny organisms, now made visible through the power of microscopy, were actually responsible for all the unwanted pooping or puking or sneezing or what have you. Like most scientific advances, though, it didn't catch on all at once.

Some argued that the bacteria that had been isolated and blamed for the disease were actually a result of the disease process and not the cause. Others argued that maybe the bacteria were to blame, but that not all people were equally susceptible—meaning the disease was only contagious under certain circumstances. One such scientist was Max Joseph von Pettenkofer.

MAX KEEPS IT CLEAN . . . MOSTLY

Max was born in Bavaria, and studied to become both a medical doctor and a chemist, but eventually wound up working largely as a hygienist. This meant that he was exceptionally good, for the time, at keeping things clean. He lived in Munich, where there happened to be outbreaks of cholera in the 1880s and 90s—so he decided to study them.

Robert Koch had recently discovered the bacterium that causes cholera, but as mentioned previously, Max didn't totally buy it. He believed that the bacteria might spread the disease, but only among patients who already lived in poor conditions, and weren't clean enough to begin with.

He had a lot of faith in his theory—we mean a whole lot of faith. So much faith in fact, that he was willing to put his money, quite literally, where his mouth was. Max considered himself a clean person, and therefore, he did not believe that he would be able to get cholera—even if the bacteria itself managed to get inside his gut.

LAST CHANCE TO BAIL:

Aww, gang, things are about to get so, so bad for the even-mildly-squeamish among you. Us. Among us. It's bad enough that we asked Teylor to follow this next part up with just the most adorable image humanly possible. I'm sorry in advance, even if you don't read it; I'm just sorry it's in a book you bought or some well-meaning, horribly misguided soul bought for you. I'm just so sorry.

Max wanted to prove his point in just the worst way humanly possible, so he . . . well, there's no delicate way of putting this: He took some diarrhea from a recently-deceased patient of his, and he mixed up history's least appealing cocktail. Once he was

Hey, look—and I'm sorry to interrupt again so soon—can we just address the fact that, because of this goofus, Syd and I had to coin the phrase "took some diarrhea"? Why does that have to exist? Like, diarrhea should never be "took." It's a linguistic impossibility really, and yet here we are.

satisfied with the consistency, or maybe the color, or perhaps he took a little time to create a garnish, he tossed back the foul concoction, and waited.

You better sit down for this one: Max got kinda sick. Hard as it may be to believe, his fecal frappé did not fill him with vim and vigor. On the other hand, Max didn't get really sick: some diarrhea, a little bit of cramping, but nothing like the more dire cases of cholera he had witnessed. He certainly didn't pass away as his donor patient had done. This was good enough for his scientific mind to prove that he didn't really get the disease, and he proclaimed it a success. Max had hypothesized that his cleanliness and socioeconomic stability would protect him, and as far as he was concerned, his theory was borne out. In reality, he probably just got a mild case, since we know that your bank account and whether or not you wash behind your ears has absolutely no impact on whether or not you get cholera. Despite this, Max is remembered as a pioneer in the field of hygiene and, in 1968, he was put on a postage stamp by East Germany.

So, that's what you get if you drink poopy water: A stamp.

DR. JOHN SNOW KNEW SOMETHING

When it came to waste disposal, 19th-century London was . . . less than ideal. Many people and businesses just dumped raw sewage into the Thames River at will—and yes, that was the same Thames River which eventually fed into public wells and water pumps. In 1854, a particularly bad outbreak of cholera was ravaging the city, and people still only had a vague understanding of how and why it was spreading.

In one of the earliest known examples of epidemiology, Dr. John Snow set out to crack the case. (Perhaps he realized that Ygritte was right; he *did* know nothing? No, no, this wasn't the *Game of Thrones* Jon Snow; we're just having some fun. This John Snow had an "H" in his first name, and was only *kind* of great with a sword.)

The miasma theory was still pretty popular among medical professionals of the time, but

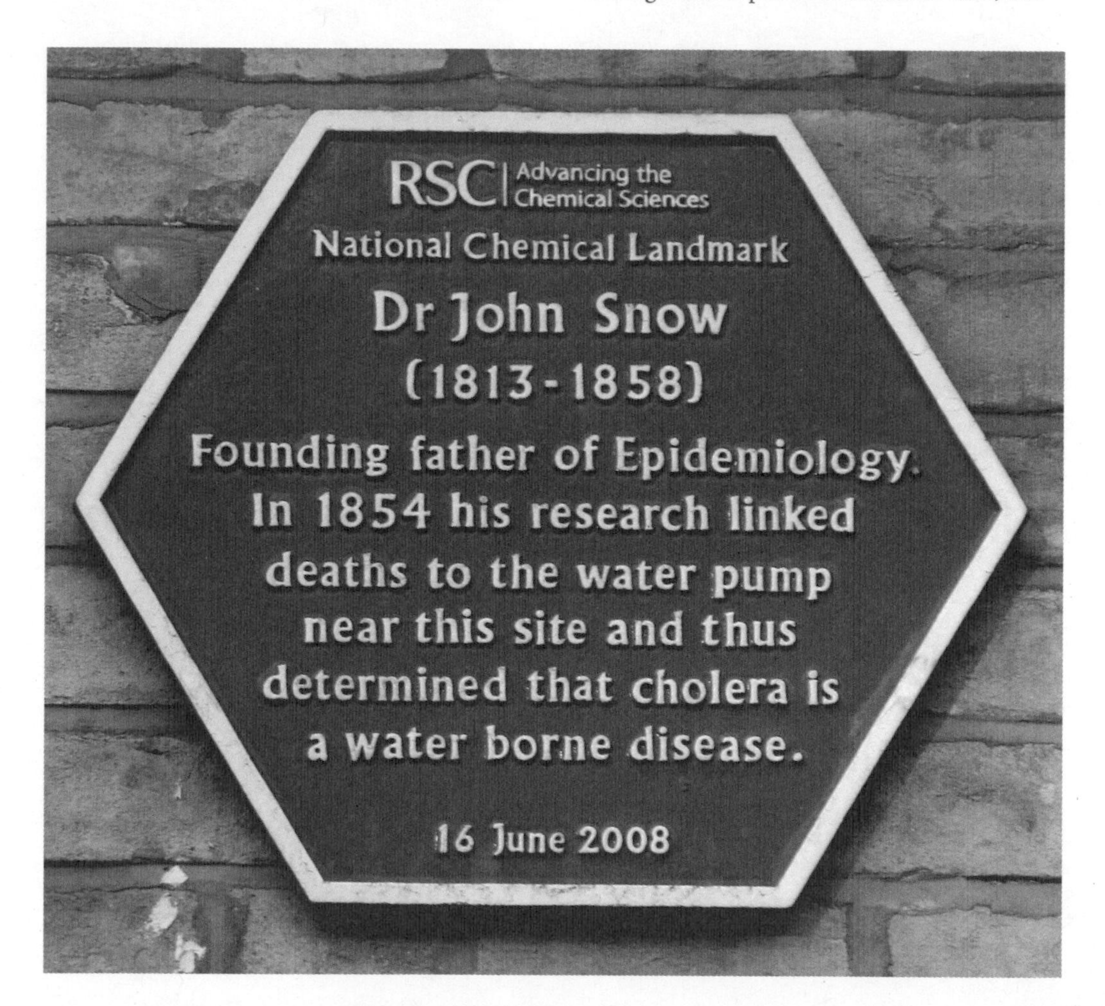

Snow was pretty sure that contaminated water was the culprit. To prove this theory, he mapped out the spread of cholera throughout the city. And indeed, he noticed a common link between many cases. The illness seemed to be clustered around a single public water pump on Broad Street. He removed the handle from the pump and cases began to decrease.

One lady almost ruined the whole thing. She lived outside the range of the pump; her niece (who also fell victim) lived even further. This really threw a wrench into his theory. However, Dr. Snow interviewed the lady's nephew, and learned that the woman had previously lived near the pump, and still preferred the taste of Broad Street water (thanks to all that tasty raw sewage, one can only surmise). So greatly did she like this robust, earthy flavor, she'd had the water bottled and brought to her by visiting family and friends on a regular basis. The niece who had fallen ill had come to visit, and the two had popped open a bottle of Cholera Water to celebrate.

Despite all this evidence, many Londoners still doubted that tainted water could have been the cause. One Reverend Whitehead set out to prove Snow wrong, based on a slightly different theory of his own. He believed that cholera was a punishment from God. He interviewed victims and published his accounts to try and contradict Dr. Snow anyway he could. Unfortunately for the Reverend, these interviews would actually reveal the cause of the initial outbreak, and further cement Snow's theory. One woman who lived on Broad Street told the good Reverend that her infant had gotten cholera before the outbreak occurred. This resulted in quite a few dirty diapers and many washings. She explained that she had been washing her baby's diapers in water that she then dumped into a cesspool . . . which was located about three feet from the Broad Street pump.

IS THIS STILL A PROBLEM TODAY?

Tragically, yes.

It wasn't until the end of the 19th century, with most of the major epidemics at last subsiding, that doctors truly began to embrace the bacteria theory. The decreasing number of outbreaks was probably related to better sanitation methods that were arising at the time. That said, currently, there are still three to five million cases of cholera each year and 100,000 to 200,000 deaths. These occur mainly in developing nations, or areas with poor sanitation or some reason for social upheaval, such as war zones, refugee camps, natural disasters, and so on.

Most heartbreakingly of all, medical advancement along with economic disparity have both made Max Joseph von Pettenkofer's theorizing about the connection between social status and cholera right—for all the wrong reasons.

MISGUIDED MEDICINE HALL OF FAME

ROBERT LISTON

1794-1847 • SCOTLAND

Surgeons have a reputation among non-surgeons as some of the most cavalier members of the medical community. But if you're looking for a real surgical swashbuckler, you'd be hard-pressed to find a more impressive one than Robert Liston, a Scottish surgeon in the first half of the 1800s. He studied at the University of Edinburgh and, by 1818, was a surgeon working at the Royal Infirmary of Edinburgh.

Liston broke plenty of ground. He became the first Professor of Clinical Surgery at University College Hospital in London in 1835, and he performed the first surgery under anesthesia in England (it had already been used in the U.S.). But if Liston has a lasting historical reputation, it's for being fast. Like, really fast.

Liston worked for a good portion of his career completely without anesthesia (remember? He was the first). Given how traumatic surgery could be in these circumstances, he figured that the most humane thing would be to get the surgery over with as quickly as possible. And, in fact, this did make sense for patient outcomes.

Well, usually.

English author Richard Gordon, who called Liston "The Fastest Knife in the West End," was a great admirer of the speedy surgeon, and he frequently wrote flowery prose about his exploits. For example:

"*He was six foot two, and operated in a bottle-green coat with wellington boots. He sprung across the blood-stained boards upon his swooning, sweating, strapped-down patient like a duelist, calling, 'Time me gentlemen, time me!' to students craning with pocket watches from the iron-railinged galleries. Everyone swore that the first flash of his knife was followed so swiftly by the rasp of saw on bone that sight and sound seemed simultaneous. To free both hands, he would clasp the bloody knife between his teeth.*"

Here, as ranked by Gordon, are four of Liston's most famous cases.

FOURTH MOST FAMOUS CASE

In just four minutes, Liston was able to remove a forty-five-pound tumor from a man's scrotum. The tumor was so large that, prior to his surgery, the unfortunate man had to carry it around in a wheelbarrow—which has to be medical history's most impressive conversation starter.

THIRD MOST FAMOUS CASE

Liston was in an argument with a colleague over whether a red tumor in a young patient's skin was an aneurism in the carotid artery (very, very bad) or a simple abscess in his skin. This next bit is, again, quoted direct from Gordon:

"*'Pooh!' Liston exclaimed impatiently. 'Whoever heard of an aneurism in one so young?' Flashing a knife from his waistcoat pocket, he lanced it. Houseman's note: 'Out leaped arterial blood, and the boy fell.' The patient died, but the artery lives in University College Hospital pathology museum, specimen No. 1256.*"

Liston is no longer with us, but we're pretty sure that even he would count this one as a loss.

SECOND MOST FAMOUS CASE

Liston once amputated a patient's leg in just two-and-a-half minutes! That's very impressive indeed, and the accomplishment is only slightly dimmed just a teensy tiny bit once you learn that Liston, in his haste, kinda sorta . . . well, he amputated the man's testicles.

History doesn't tell us whether or not the patient specified that he didn't want his testicles amputated in the procedure, so assigning blame feels like a toss-up on this one.

LISTON'S MOST FAMOUS CASE

We've all had a rough day at work, but you'd be hard-pressed to compete with Robert Liston's terrible, horrible, no-good, very bad day.

It again begins with Liston amputating a leg in two-and-a-half minutes (maybe slow down, my dude). The patient died in the hospital from gangrene, but that happened a lot in the days before antibiotics. No, the really impressive thing is that Liston also amputated the fingers of an assistant who was helping to restrain the patient at the time . . . and that assistant ended up also dying of gangrene.

It gets worse.

During the procedure, Liston also managed to nick a doctor observing the surgery who, according to Gordon, "who was so terrified that the knife had pierced his vitals he dropped dead from fright."

So here's to you, Dr. Robert Liston, perhaps the only person in recorded history who conducted a surgery with a 300-percent mortality rate. You might not have always cut the right stuff off, Doc, but at the very least, you never made your patients late for anything.

Unless they died. Obviously.

Urine Luck!

Welcome to Pee-Pee's Playhouse.

Urine has long been known as a diagnostic tool for a number of conditions. Perhaps you're thinking doctors throughout history have left it right there, satisfied with its useful properties, and with no desire to turn it into a magical panacea. If so, you haven't been paying attention, or you're skipping around. Shame on you.

There is nothing a physician values quite as much as a decent sample of their patient's bodily secretions to examine. So much can be learned about our health from a properly collected and analyzed specimen, neatly deposited into a plastic cup. Surely, the most popular fluid throughout history must be the humble urine sample. Whether smelling it, tasting it, chemically separating its components, or simply gazing at it in bright sunlight, urine has captured the imagination of doctors since early history.

Sydnee started the chapter like this, seriously, and I was tempted to juice it up with my own erudite wit (which is to say, lots and lots of dick jokes). But honestly, an earnest love note to pee-pee was just so wonderfully, adorably Sydnee that I have left it untouched so that it may be captured for time immemorial.

In case you were curious what urine is made of, we're happy to let you in on one of medicine's best-kept secrets: It's water. Well, it's mostly water, but with a sprinkle of electrolytes in there, such as sodium, potassium, chloride—

Oh, so it's just like Gatorade!

—as well as waste products including urea, and random other things that your body has carefully filtered out through your kidneys.

Okay, well that's less like Gatorade. At least it's not like any particular flavor I've ever had. Did I miss Arctic Urine Blast?

Despite what you might guess considering the presence of waste products, urine is actually sterile. Well, it's technically sterile until it travels through the urethra, which is . . . so not sterile.

The earliest writing we've found about any medical condition is from around 4000 BCE, discussing urinary tract diseases. The ancient Babylonians and Sumerians would record their findings on clay tablets, which is how, millennia later, we can still benefit from this treatment.

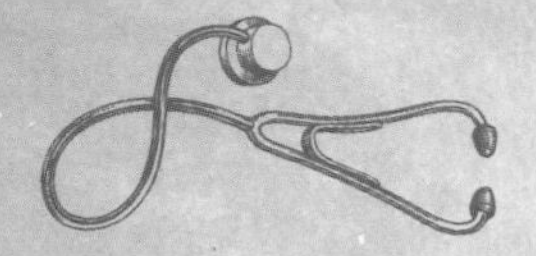

Sydnee's Fun Medical Facts

The kidneys are truly exquisite in their design and function. They help regulate your blood pressure, the level of acidity in your blood, your fluid balance, and your electrolytes, and they are responsible for removing waste products from your body. Because they need to perform all of these vital functions your whole life, they are actually designed to fail and regenerate, instead of continuing to take damage and decrease in function. Like a pair of tiny little bean-shaped Timelords, your kidneys will die and regenerate if taxed too badly, so that you emerge with fully functional organs as opposed to continually declining over time.

"If a man's urine constantly drips and he is not able to hold it back, his bladder swells and he is full of wind, his urine duct is full of blisters; in order to cure him, crush some puquttu seed in pressed oil, and [blow] it into his penis through a bronze tube."

PEE THROUGHOUT THE AGES

Hippocrates relied heavily on the prognostic abilities of pee, and thought that the urinary system was the most important in the body for diagnosis. Discolored, oddly-scented pee was like the human body's "Check Engine" light, and Hippocrates was determined to check the . . . umm . . . maybe it was more like the human body's dipstick? Listen, okay, we're not car people. Bad pee meant sick person, and the type of bad could tell you the type of sick. How's that?

This concept was taken to the extreme in the Middle Ages, when it was thought that all diseases could be diagnosed by examining urine. This practice, known as uroscopy, was so relied upon that some doctors would diagnose you strictly by examining a bottle of your urine.

Urine was also used to tell if someone was a witch: You simply put the suspect pee into a bottle and add a variety of metal objects, and then place a cork in the top and wait. If the cork stays put, they are probably cool. If it pops out, your patient is a witch. (Also, what on Earth is with their urine?)

Little known fact, medieval doctors would often have everyone at a party pee into cups and then use the power of uroscopy to match each guest to their urine. Lesser-known fact: Medieval doctors weren't invited to many parties.

NOT IN GOOD TASTE

Up to the early modern era, a thorough sensory evaluation of urine was thought to be essential for diagnosis—including taste. In his 1911 book, *The Evolution of Urine Analysis*, Sir Henry Solomon included a section translated from Sanskrit, on Ayurvedic "tasting notes" for urine. They had twenty different varietals indicating a bevy of different maladies. Here are a few of our favorite entries, and possible modern diagnoses.

***Iksumeha* (cane-sugar juice urine):** The urine is very sweet, cold, sticky, and opaque, like the juice of cane sugar. This may indicate diabetes.

***Sandrameha* (thick, fluid urine):** The urine becomes thick after standing some time. This is often due to excessive mucus in urine, which can signal anything from infection to bladder cancer.

***Surameha* (urine like brandy):** The urine is clear above and turbid below. This has a high correlation with too much calcium or phosphate in the body, being passed in your urine.

***Lalameha* (frothy urine):** The urine has threads and is passed in small quantities. Today we know this as albuminuria: proteins in the urine from kidney damage, or too much salt intake.

***Pistameha* (floury white urine):** When a patient passes this variety of urine, the hair over the body becomes erect, and urine looks as though mixed with flour. This may be chyluria—fats and lymph in urine from blocked ducts.

Anybody else getting thirsty? No? Us neither.

The Ayurvedic system was a great way to find the most refreshing local homebrew urine, but it wasn't scientific (or circular) enough. Luckily, later physicians and researchers had—we're so excited we get to type this—the Urine Flavor Wheel.

You know, medical terminology can often be so confusing for a layman like me and (maybe) you, that it's really refreshing to have a bit of jargon in this book that is really 100-percent what it sounds like.

You didn't need to be a doctor to use this tool; all you had to do was look on the wheel for the color, texture, and flavor of the urine in question, and you'll know what ails the patient. Well, you usually wouldn't, because it was the 1400s or whatever, but sometimes—believe it or not—you totally would.

That's right—the really wild thing about the Urine Flavor Wheel is that sometimes it was totally accurate. For example, a diabetic patient might have excess glucose in their urine if the glucose level in their blood is running high. So using our handy-dandy Urine Flavor Wheel, we can deduce that a patient has diabetes if their urine is astringent, sweet, and sharp. It's just like tasting wine! If you didn't want to drink pee, you could also just pour some on the ground and see if ants came. Ants like sweet things, so they could do the work for you in the case of diabetes.

THE PEE TREATMENT

The idea of using urine to diagnose illness, no matter how strange the method, is something we are very familiar with. Using urine to *treat* disease, though, is a lot less common these days. Our old friend Pliny the Elder, for example, advised using urine for "sores, burns, affections of the anus, chaps, and scorpion stings." He also made special mention that you could mix stale urine with ash, and apply it to your baby's bottom for diaper rash. Or you could brush your teeth with it. Of course, Pliny also advocated spitting into your urine immediately after voiding to prevent being cursed.

One popular remedy from ancient Greece advised treating a fever by boiling an egg in your patient's urine, and burying it in an anthill. You have to keep an eye on the anthill, though, because the disease won't wash away until it does.

Now we'll jump forward a millennium or so to the 16th century. Ambroise Paré was an incredibly progressive and influential French surgeon of the era who really did come up with a lot of revolutionary and correct ideas about wound healing and postoperative care. He also advised that you should keep urine all night in a barber's basin, then wash your eyelids with it to relieve itching. His French colleagues also advised soaking your stockings in urine, and wrapping them around your neck to treat a sore throat.

GOLDEN SHOWERS OF HEALTH

There are a few accounts sprinkled through history of urine-based treatments. For example, in 1550, we read that Italian doctor Leonardo Fioravanti urinated on a man's nose after it had been sliced off in a dispute and then sewn back on (in an attempt at healing, not to add insult to injury.) Henry VIII's surgeon, Thomas Vicary, advised washing all battle wounds in urine.

Robert Boyle, a 17th Century Irish chemist, advised that patients drink some of their own urine every morning to maintain good health. He only advocated for a moderate amount, and he noted that warm pee was best. Surely we can all agree on that.

SYDNEE There is a condition called "porphyria" in which your body doesn't properly turn substances called "porphyrins" into heme, part of hemoglobin, which is how your blood carries oxygen to your body cells. The buildup of all these excess porphyrins can cause a lot of symptoms, among them being urine that turns purple when exposed to UV light. The name porphyria actually comes from the Greek for purple. So if you suspect someone may have this condition, you could just put their pee in your windowsill for the weekend and see what happens! Not that I've ever done that before . . .

JUSTIN She totally has.

In 1666, urine was used against the plague, but to be fair, the plague was really bad, and everybody tried everything against it, so lots of weird treatments were attempted. This same excuse can't be used for the vast amounts of children's urine that were boiled down into essence of urine for use in reviving those who had been struck by "the vapours." Query: Isn't this sort of legit though, in that ammonia is still used as "smelling salts"? Available on Amazon today (I checked to be sure I wasn't nuts). I mean, not made from childrens' urine probably, but still.

LIVE-STREAMING THE MODERN ERA

In World War I, word went out that you could soak a cloth in your own urine, and hold it over your face for use as a gas mask. Ammonia in the urine was thought to counteract the chlorine in the gas. If you know a little chemistry, though, you may already know that this combination would instead become toxic.

In many parts of the world, nontraditional and folk remedies still use urine. A search using the term "urine therapy" (we recommend that you wait until after lunch to do this) turns up thousands of pages where people recommend drinking, bathing in, or massaging yourself with urine to treat all sorts of ills, from epilepsy to cancer. Many American pioneers believed in the application of pee to treat earache, a treatment Sydnee still hears advocated today.

Okay, Sydnee isn't allowed to share her medical opinions, lest you think she's communicating to you directly, and you sue her for malpractice or something. Listen, I'm sketchy on the legal details. But I'm a layman, and I'm gonna come down pretty hard on "you shouldn't put pee in your ear."

HOW IS URINE USED TODAY?

Tasting urine has fallen out of favor, but using our sense of sight to evaluate urine is still quite practical. Unusually colored urine is still a great reason to visit your physician, though there are some common reasons that pee might look a little off. Here's a handy color-coded cheat sheet although only your doctor can tell for sure what's going on with your urine.

DARK YELLOW Dehydration

ORANGE B vitamins, medications, jaundice

RED Blood

PINK Beets

GREEN Asparagus

BLUE Any of a number of medications

In theory, smelling urine could still be a helpful diagnostic, but most health care professionals probably just prefer a urinalysis. A really unpleasant smell could mean infection, a sweet smell may signal diabetes, and some foods, such as asparagus, leave telltale scents.

Through the years, drinking urine has been advocated for fertility; to stimulate sexuality; for fevers, yeast infections, oral infections, diabetes, and bladder problems; to help break down blood clots; or to treat HIV or cancer. (It should go without saying that the American Cancer Society advises against drinking urine to prevent cancer.) Before we leave you: There's an episode of *Friends* where Chandler pees on Monica after she's stung by a jellyfish to ease the pain. This does not work—do not pee on your friends.

MIRACULOUS UNIVERSAL CURE-ALL RADIUM

Radiation is an invaluable force when you want to eat a frozen burrito in the next three minutes, or if you need to find a particularly sneaky cavity. Did you know we even use it to sterilize food? It's true!

But the wrong amount of or kind of radiation? Well, that can be very super-duper not great. You probably didn't need us to tell you that; there is, after all, a big scary symbol created for the express purpose of alerting us highly permeable humans that dangerous radiation is present.

Of course, that didn't stop a few highly unscrupulous or misguided individuals from trying to harness the power of radium to cure (or, you know, deffo not cure) illness.

It's worth noting that when radiation's use as a panacea was at its peak, we had no idea how bad it was for you. Consider this bit from the manual of the Radium-ore Revigator, a device from 1928:

"Radio-activity is not a medicine or drug, but a natural element of water, and that since practically all spring and well water that Nature herself gives for drinking purposes contain this highly effective beneficial element, it is but common sense to restore it to water that has lost it just as we restore oxygen to a stuffy room by opening a window."

USED TO TREAT:

What *couldn't* radiation cure? Most things, but that didn't stop folks from trying. Here are just a few of the supposed uses.

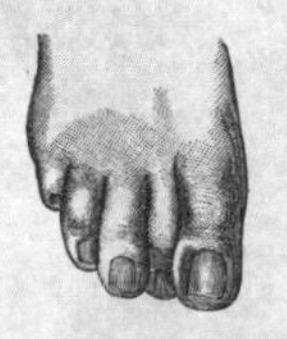

ARTHRITIS AND GOUT

Low levels of radiation were discovered in therapeutic hot springs, but how to get all that delicious radiation at home? Enter radium bath salts, designed to be dumped in the tub for the treatment of arthritis and gout.

IMPOTENCE

We have Harvard-dropout William J. A. Bailey to thank for Radithor, a blend of radium, mesothorium, and drinking water. Bailey (who frequently claimed to be a doctor) designed it as a general health enhancer, but with specific focus on treating impotence. Wealthy socialite Eben Byers swore by the stuff and sucked down three bottles daily—right up until the day his jaw fell off. Byers' eventual death from radium poisoning triggered a cease-and-desist for Bailey from the Federal Trade Commission that soon put him out of business. (Things worked out okay for Bailey in the end; during World War II, he managed the electronic division of IBM.)

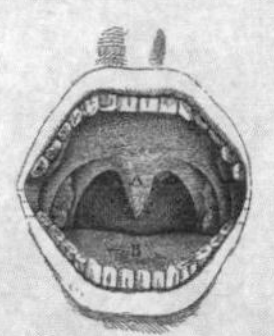

SHINY WHITE TEETH

German-made Doramad boasted a secret ingredient Crest and Colgate never had the guts to include: thorium for antibacterial action! Fun fact: In 1944, the Allies were thrown into a panic when they noted a massive shipment of thorium into Germany. They assumed shipment signaled the Axis had created atomic weaponry. In reality? The thorium was for Doramad.

SAFER CIGARETTES

The NICO Clean Tobacco Card was made in Japan, and exported to America. It was a metal plate treated with uranium that was supposed to be slid into a pack of cigarettes. In twenty minutes, the plate would (its creators claimed) lower the cigarettes' tar and nicotine content by 17 percent. "Here, have one of my cigarettes," the Clean Plate user could confidently offer. "They're irradiated for your health!"

FLATULENCE AND SENILITY

The Radium Ore Revigator was perfect in its simplicity: a pot covered in radioactive material that would transform your drinking water into straight-up poison. Generally, the radioactive water was supposed to "create cellular energy and remove cellular poisons," but some of the specific maladies to be treated included flatulence, arthritis, and senility. In the modern era, scientists found that while a vintage Revigator didn't add enough radon to treated water to be dangerous, it *did* infuse it with detectable levels of arsenic, lead, and uranium. The Radium Ore Revigator: At Least It Won't Kill You In the Specific Way We Advertise.

Humorism

Sure, all the other chapters have been gut-busters, but this one is guaranteed to leave a stitch in your side.

Get your funny bone ready for a hilarious tour of all the zaniest ways we've tried to use great jokes to treat all manner of illness as we prove laughter truly is the best medicine. Also, we're probably gonna talk about Patch Adams, so it's time to get psyched!

. . . I was alerted just now by my wife that this chapter has nothing to do with humor, and is about something else entirely. I deeply regret this error, caused by a sadly unavoidable occurrence of me writing the opener before I read the chapter. I would also like to apologize for my unwillingness to edit the above text to reflect the actual content of this chapter. This situation was also unavoidable, as it was a lot of words ago, and I am a very busy man.

The idea that every individual's health, wellbeing, and even personality are determined by the balance of essential bodily fluids, or humors, is one of the oldest in history. There are indications that these beliefs got their start in ancient Egypt, or maybe Mesopotamia, but it took the ancient Greeks to turn those ideas into a comprehensive system of diagnosis and treatment. This system was incredibly long-lived; in fact, it formed the foundation of most medical thought, all throughout Europe until the 19th century. On the Indian subcontinent, Ayurvedic medicine took a similar approach for millennia, and is followed by many today.

It would be natural to assume that an idea that spanned both time and geography to such an extent must be correct. However, as devoted students of medical history, we can tell you that being wrong never halted a concept if the doctor behind it just said it loudly and with confidence. Alongside "do no harm," those old-timey doctors probably should have embraced the motto "always certain, but rarely right."

ANYTHING THIS SIMPLE HAS TO BE TRUE . . . RIGHT?

One very attractive facet of humorism is its simplicity. All you really need to know is that there are four fluids within the human body that must be balanced in order to maintain physical and mental health. These fluids, or humors, are created in the liver as the products of digestion, and then stored in various organs. You're probably wondering what the four humors are, but please, be patient. First, we've got to lay a little blame.

The first iterations of medical humorism may have come from Hippocrates, or his son-in-law, Polybus of Cos, but it was brought to real prominence and further defined by the Roman physician and philosopher, Galen of Pergamon. The actual nature of the four humors and their individual significance was a product of Galen's many writings.

So wait, Hippocrates or maybe Polybus didn't even say what the four humors were? He was just like, "Hey yo, check this out: There are four liquids running everything. They're in your tummy or whatever. I don't know what they are yet, but I'm totally right on this one. Trust me on it."

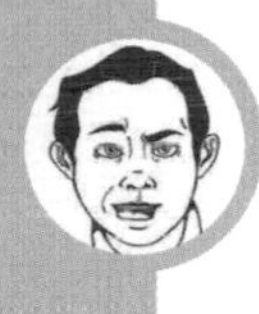

It's pretty clear that Galen approached humorism as a theory, attempting in his writings to flesh it out with detail based on his understanding of the human body. However, once Galen was no longer around to properly explain things, many of his followers began to take his idea as absolute fact. This galvanized the humoral theory and would make it unshakable for many years to come.

. . . OR NOT. POSSIBLY NOT.

In fact, the theory of humorism remained much the same for over a thousand years, and in many traditions. Each humor was thought to have healing or harmful properties based on their associated temperature and moisture content. Balancing out the humors usually involved

THE FOUR HUMOURS

Okay, so, you've waited long enough!
Here are the four humors. Enjoy!

PHLEGM
Cold and Wet
Stored in the brain and lungs
Helps the body purify itself
Element: Water
Season: Winter
Age: Old age
Temperament: Phlegmatic

BLACK BILE
Cold and Dry
Stored in the gallbladder
Used by the body to make bones, teeth, and connective tissue
Element: Earth
Season: Autumn
Age: Adulthood
Temperament: Melancholic

YELLOW BILE
Warm and Dry
Stored in the spleen
Helps with digestion
Element: Fire
Season: Summer
Age: Youth
Temperament: Choleric

BLOOD
Warm and Moist
Stored in the liver
Contains a vital essence
Element: Air
Season: Spring
Age: Infancy
Temperament: Sanguine

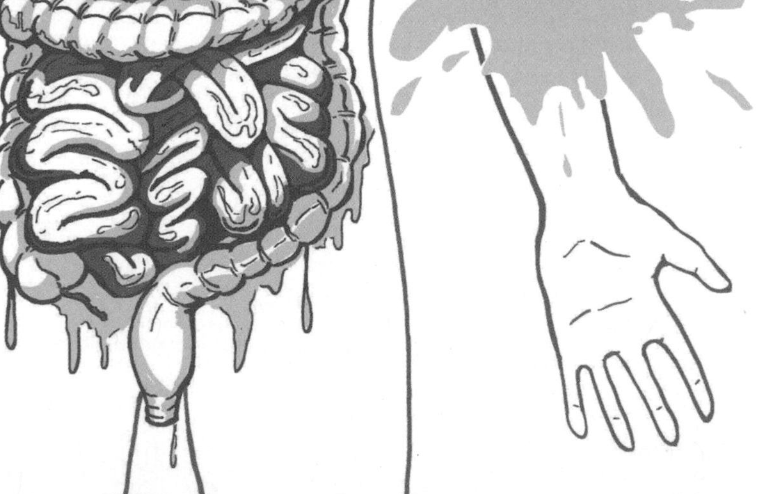

working to improve or counter these various properties, based on the patient's condition.

A LITTLE HARD TO SWALLOW

Before we go further, none of this will make sense unless you know that we really didn't understand digestion at all for much of history. The ancient Greeks believed that digestion happened in four totally separate phases.

Phase one was supposed to take place in the stomach, resulting in formation of stool (which is to say, poop, in case that wasn't clear.)

Phase two was a big deal—it happened in the liver, and produced all four of the humors.

Phase three happened in the blood vessels, producing urine and sweat.

Phase four was your body's chance to get rid of any last "abnormal humors," passing them from the body as ash.

In phase two, the humors were supposedly created in order of importance. First nutrient-rich blood, followed by a whole lot of phlegm to keep the respiratory system running smoothly. Then, a little yellow bile to keep the digestion going right, and finally, a very small amount of black bile for our bones.

You may be wondering why no one just investigated this theory by simply taking a peek under the hood of a cadaver and seeing if the humors were, you know, there. It seems obvious to us super-smart modern folks that a simple examination would have proven that humorism was not actually based in fact, and instead just made up from whole cloth.

If you believe the theory Swedish physiologist Robin Fahraeus put forth in the 1920s, they totally did. You see, if you draw blood from a vein and then let it sit out in a container for a bit, it'll separate out into layers. A dark clot will form at the bottom, followed by a layer of red blood cells above that, a whitish layer of white blood cells above that, and finally, a layer of yellowish serum at the top. There are your humors.

It's a tempting theory that would account for the sustained belief in the existence of the four humors. However, since the creation and spread of humorism predates blood draws and test tubes, it may be a bit of a stretch.

DON'T LOSE YOUR BALANCE

Each humor had specific qualities associated with it, both good and bad. The key to good health was thought to be keeping them in proportion. Of course, the right balance was thought to be different for every person—that extra helping of black bile that makes me so thoughtful and deep might just make you sad and lonely. Achieving the correct balance was partially just trial and error, and even if you did hit that perfect fluid equilibrium, you could throw it off again with the wrong diet or exercise routine.

The essence of medical care in the age of the four humors, then, pretty much boiled down to eating and drinking things and taking medications that would crank selected humors up or down in your body. Your prescription would be based on the temperature and moisture of the humor that was out of balance. So for example, eating cold foods were thought to make you produce extra phlegm, while hot foods would get you a nice healthy dose of yellow bile.

All that tinkering with your diet a little too low-key? Good news. Common wisdom was that the best way to get rid of whatever excess humor was getting your down was to expel it from the body as efficiently as possible. Bloodletting, obviously, was the best way to relieve excess blood. Diuretics would get rid of excess phlegm, and laxatives and emetics (stuff that makes you puke) could help clear out extra black or yellow bile. As a result of this theory and manner of treatment, a physician was very much a dietician as well, telling you what to or what not to eat based on your humor levels.

SYDNEE Not only was this treatment easy to explain to patients, it played to one of the major strengths of early physicians that has persisted throughout medical history. Doctors know how to make you poop and puke. Even before we understood why or how or if we should do these things, we definitely had a full armory of noxious herbs and chemicals that would clean you out from one end or the other. As an added bonus, the patient never needed to be convinced that the prescription they were given was working because they always knew. God help them, they knew.

JUSTIN . . . Wow, Syd. It's really inspiring company you've thrown your lot in with. I always wondered when your class had to recite the Hippocratic Oath, what the line "And I shall take it to the limits of nastiness, twenty-four seven" was in reference to, but I think I'm starting to get it.

SYDNEE You can laugh all you want, but my particular set of skills may just save your life someday. Or at least your weekend.

JUSTIN Because I'd be . . .

SYDNEE Constipated, right.

It wasn't all diet, exercise, and leeches. Sometimes, things got really serious. If you had the plague, for example, you had too much of all of your humors in general, so arsenic was good for clearing out all that excess.

Arsenic would not be good for that. Or for anything, really. Unless, by clearing out excess humors, you mean dying. Arsenic would actually be good for that.

Remember how we mentioned above that the humors weren't just supposed to affect your health, but your personality as well? Changes in humors were thought to explain changes in temperament or moods. It was thought that everyone was sort of guided by one humor or another and that it dominated the way they behaved and interacted with other people. But which excesses led to which temperaments?

Here's a way to remember the humors, by the third-tier *Simpsons* character they remind us most of.

Black bile Too much of this yucky stuff supposedly caused depression, as well as melancholic temperament. That's sad sack bartender Moe Szyslak for sure.

Yellow bile An overabundance of yellow bile was said to cause anger or derangement, and a choleric, irritable temperament—clearly what drove Sideshow Bob to a life of crime.

Phlegm Too much phlegm caused you to be apathetic, which is pretty much Mrs. Krabappel all over.

Blood Probably the only good humoral side effect on our list, too much blood led to a carefree, happy sanguine temperament. After two hours of contentious debate, we're declaring it a tie between Disco Stu and Doctor Nick Riviera, in the hopes of preserving our marriage.

YOUNG, DUMB, AND FULL OF BILE

In the diagram above, you'll see that each humor is also associated with a stage of life. The thinking went that, while each individual may have a guiding humor through all of their days, different phases of life were connected to each of the humors as well. You start out sanguine when you are a child, and you are filled with a lot of blood. The teenage and young-adult years are full of passion and upheaval from your new choleric temperament. When you become an adult, you start to worry about stuff, and you are depressed and melancholic. (We can vouch for this one.) Then you get old and kinda say "screw it." You are at peace with it all and have become phlegmatic.

These "medical" definitions bled into works of popular culture at the time. It wasn't uncommon for theatrical characters to be introduced as having temperaments related to humors. In *The Taming of the Shrew*, for example, Katherine is kept from eating a calf foot because it's too "choleric" a food, and she's already grouchy.

DO WE STILL DO THIS TODAY?

Short answer? No. These concepts are deeply woven into European languages and literature, but medical science has moved on. Notably, some alternative treatments touted even today have their roots in the humoral system. A good example of this is "cupping." While Olympic athletes may do it because they believe it increases blood flow to certain tissues, it was practiced through much of history because of its perceived ability to migrate the humors around in the body.

THE DOCTOR IS IN

It's time to take another break from rubbernecking at the flaming wreckage on the medical history highway to answer real questions from real listeners of *Sawbones*, a real podcast.

I know that apple seeds contain cyanide, but how many would you have to eat for them to be fatal?

Sydnee: Depending on what kind of apple we're talking about, you'd need to eat between 150 to several thousand apple seeds. Oh, and you'd have to crush and chew every one. Gross. By the way, they don't contain cyanide, but rather a family of chemical called "cyanogenic glycosides" that are transformed into cyanide in your gut. This is not just apples, by the way: cherry pits, peach pits, and even lima beans all contain trace amounts of cyanogenic glycosides.

Justin: Good news, kids: you finally have a great excuse not to eat lima beans!

Why, for some women, does your hair change after pregnancy?

Sydnee: The general answer is "hormones," but the main factor is estrogen. This can cause your hair to grow faster, and change the texture, but also enter what's called the "resting phase," during which your hair doesn't fall out. The trade off is about three months after delivery, your hormones revert and then your hair falls out.

Justin: Fall out? Not if I pull it out myself first! Right, because kids? Where my other parents at? Anybody?

I'm an avid blood donor, and I've recently been thinking: If a baby got my blood, is it possible that a blood paternity test could come back as a match if ran against my own blood, or is that not how any of this works?

Sydnee: You are exactly right in that this is not how any of this works. Modern paternity tests largely rely on analyzing DNA from cells collected by swabbing the inside of your cheek. Older tests relied on blood, but only the white blood cells, because red blood cells (which are what you're actually donating) don't contain any markers that can be used to determine their parentage.

Justin: I was trying to come up with something funny to write here, but honestly, I'm actually just too excited by that *Sawbones* crossoverwith *Maury Povich* that I've always dreamed of, finally coming to fruition.

When I was a kid, I had an allergy to apples. Brilliant ten-year-old that I was, I decided to beat my allergy by eating one every day and just enduring it. A few months later, my allergy was gone. Is there any merit to this, or was I just lucky?

Sydnee: Let me preface this answer by first saying, in no uncertain terms, DO NOT DO THIS. While this story, fortunately, has a happy ending, this is not a risk I would ever advise taking. That being said, this is actually similar to how allergy shots work. You receive a series of injections in this treatment, with a gradually escalating dose of the allergen itself. In this way, you develop a tolerance of sorts to the allergen.

Justin: People used to scoff when I would rub a cat on my eyes and nose for fifteen minutes every day. But it looks like I wasn't so dumb, huh? Even better, Parker Purrsey can rest easy in the Great Beyond knowing that she got to achieve something truly important during her time on Earth.

My husband will sometimes gorge himself and eat so much he has to lie down, but when there's dessert, he's like "Heck yeah" because he has a theory that "Dessert fills in the cracks," like the cracks between the food he just ate. That's ridiculous right?

Sydnee: Right.

Justin: Finally, one even I could answer!

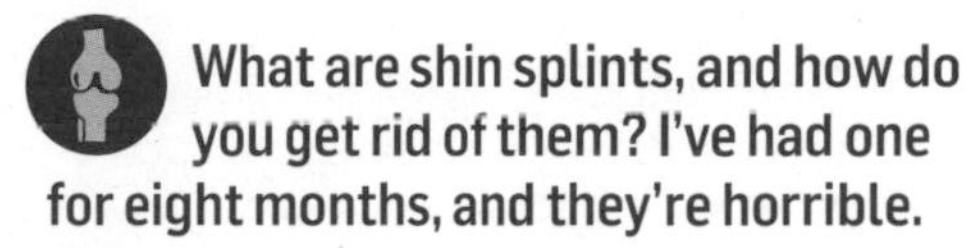

What are shin splints, and how do you get rid of them? I've had one for eight months, and they're horrible.

Sydnee: Shin splints, or Medial Tibial Stress Syndrome, really just means that you have pain along the inside edge of your tibia. This is related to inflammation of the tendons, muscles, or tissues that surround the bone, and is typically the result of overuse. It is extremely common in runners. Most patients will respond to a combination of rest, ice, compression, stretching, and maybe an anti-inflammatory medication. If this isn't helping, though, you should go talk with your doctor. Sometimes, what you're thinking might be a shin splint is actually a very small stress fracture, so it's best to get it checked out from the start.

Justin: Boom, take that, runners! Now who's the healthy one? Boy, this has been a really vindicating segment so far, Syd!

What can a doctor determine about a patient after observing a fart?

Sydnee: Contrary to what you may think, a fart is actually not a particularly helpful diagnostic tool for a physician. I guess a very smelly fart may clue you in to a high-protein diet or an abundance of gas may lead you to believe a patient is eating a lot of beans, but beyond these simple dietary clues, there really isn't much to learn from flatulence. Also, I would personally prefer to just ask about these things; no demonstration required.

Justin: . . . I think I finally understand why the producers of *ER* never wanted to produce that spec script I sent them.

The Straight Poop

The number two thing you can do for your health. Aww, thanks folks; we'll be here all week.

People don't really use the term "the straight poop" in casual conversation anymore, so in the rare cases it is employed, it's usually a double entendre to indicate that feces will be discussed shortly. We have presented this observation here for reasons that will become immediately apparent.

I thought I would just start this chapter out as disturbingly as possible and get everyone on board with the fact that things are about to get real gross.

Human beings have been using poop as medical therapy for a really long time. How long? Well, some Chinese medical writings from the 4th century discuss the practice of ingesting a solution made with fecal material as a treatment for diarrhea and food poisoning—a practice that we, at our kindest, would describe as "counterintuitive." Li Shizhen, the great physician and naturalist of the Ming Dynasty who wrote the definitive treatise on Chinese medicine of the time, the *Compendium of Materia Medica*, also advised the use of "yellow soup" or "golden juice" for a lot of abdominal issues. I can only assume that these euphemisms were an attempt at making the treatment sound more appetizing. However, considering the recipe consisted of fresh, dry, or fermented stool mixed with water, there probably wasn't any way to do that.

Syd, I'm sorry. I . . . I don't think I can with this one. I don't want to abandon you to write the book on your own, but I'm doing exactly that, I guess. I'll check back in a bit to see if things have, uh . . . cleared up.

You might be relieved to hear that the ancient Egyptians didn't eat any poop (that we know of), but that relief is going to be short-lived, because I'm about to tell you about ancient Egyptian birth control. In the Kahun Papyrus, which dates from 1850 BCE, and is the oldest known medical text in the world today, advice is given on creating a vaginal tablet of crocodile dung, combined with fermented dough or honey and sodium carbonate, in order to prevent pregnancy. This was probably, at least in part, connected to the fact that crocodiles were associated with the god Sobek, a deity whose worship was related to fertility. Aside from the fact that this is a terrible idea, and you should not interact with crocodiles or their dung any more than absolutely necessary, it was probably ineffective as well.

That's right; we said "probably." Bear with us: In this highly unpleasant thought experiment, one must acknowledge the fact that croc poop is alkaline. As such, it may have actually created a more hospitable environment for conception. Hypothetically speaking, anyway.

This is implied for the entire book, but just to be doubly sure, please don't use crocodile poop for anything related to matters of conception—or contraception, either, for that matter.

FLUSHING OUT THE TRUTH

Just as we learned in elementary school, everybody poops. And humans, being the endlessly creative and resourceful species that we are, have found use for an awful lot of that diverse dung beyond fertilizing our gardens, burning it for fuel, and making really disgusting diaphragms. Here are just a few of my favorite examples.

Ancient Hindu texts advise cow dung for wound cleaning and for a facial scrub. You could also consider trying dried dog poop as a remedy for a sore throat. Pliny the Elder, not to be outdone, advised the use of rabbit pellets for a stubborn cough.

As we just discussed, the humoral theory of medicine dominated medical treatment from the time of Hippocrates onward (you're reading this book in order, right? If not, just nod and smile, and pretend you know what we're talking about). In his attempts to understand and balance the humors, Galen found feces to be useful both

externally and internally for all manner of things. The reasoning for this was simple, if disgusting. Human excrement could make you puke, which was kind of like a hard restart to rebalance those humors. Simple, yet gross.

In 1696, a German physician named Christian Franz Paullini wrote the *Salutary Filth-Pharmacy*, which consisted of a wide range of prescriptions for various ailments, based on bodily excretions. This included urine, earwax, menstrual blood (not his), and of course, the use of human poop as a treatment for dysentery.

"Whoever disrespects feces," he wrote, "disrespects his origin." By the way, if you're looking for a new inspirational tattoo to get, we think you've found it.

A CRAPPY SURPRISE

The most creative and, certainly, historically significant use of poop dates back to World War II. German soldiers stationed in North Africa observed that the region's Bedouin people ate fresh, warm camel dung as a treatment for dysentery. Eventually, those soldiers picked up on the stinky life-hack and started eating fresh dung themselves.

We have to assume this treatment was at least somewhat effective, because eventually camel droppings became good-luck symbols for the German soldiers. In the same spirit that a basketball player may reach up and smack a door frame on their way to the court, so too would German soldiers try to run their tanks over the piles of camel dung for good luck.

I leave for a couple paragraphs, and I almost miss tanks driving over dookie? The thirten-year-old boy that forever lives inside my heart would have been so disappointed! "Tanks driving over dookie" ranks just below "He-Man slamming a Capri-Sun" in his extensive list titled "Rad Stuff I Want to See."

This odd practice did not escape the notice of the Allies. Taking full advantage of this opportunity, the Allied soldiers started disguising land mines as camel dung in order to blow up those tanks, which is about the least lucky thing that could have happened. It would be like the basketball player smacked the door frame and caused several stories of the stadium to collapse on him.

It only took a few run-ins with this explosive dung for the German forces to realize that their luck had run out, and to stop the practice of driving over dung. Unless, of course, the dung had already been clearly smashed by tank tracks, in which case it had been proven safe and could be run over again. Once this fact was recognized by the Allied soldiers, they started making land mines that looked like dung that had already been run over by a tank. At some point, we can only assume, soldiers realized that as fun as it was, running tanks over camel poop just wasn't worth the risk.

DISCOVERING THE SECRET INGREDIENT

Eventually, German scientists figured out the key component of the camel dung that stopped the dysentery: Gut bacteria, such as the kind found in the camel caca that help with the digestive process, can also be very helpful in treating diarrheal illness. While you can certainly obtain these bacteria from actually ingesting fecal material (we're not here to judge) all you really need are the bacteria themselves.

So, the Germans figured out how to isolate the bacteria in camel dung that was so helpful, allowing soldiers to just take the bacteria as a supplement instead of eating the actual dung. We can't imagine many memos in military history that would have been met with more jubilation.

These helpful bacteria are, of course, not only found in camels' digestive tracts. Both horse and cow manure have been used to filter water due to the presence of bacteria that can remove impurities. We'd probably reach for a nice Evian first, but hey, desperate times . . .

DO LIKE THE ANIMALS DOO

We generally limit ourselves to discussing human medicine only, but with this specific topic, allow us a brief foray into veterinary medicine. (There's a good reason, we promise.) Good gut bacteria is the reason that some other members of the animal kingdom, such as hippos, koalas, pandas, and elephants, eat their mothers' dung: The babies are born with sterile guts, which prevents them from

performing all the digestive functions necessary for survival. Eating their mom's stool will populate their guts with that good bacteria.

This idea has been employed in veterinary medicine since around the 17th century in Italy. Ruminant material from the stomach of one cow or sheep is transferred to another, to help regulate and correct their gut flora; this process is known as "transfaunation."

It seems fitting that, sooner or later, human beings would want to fully embrace the good bacteria theory, and start treating poop with more poop. In 1910, a doctor was documented in the *Journal of Advanced Therapeutics* injecting *bacillus* bacteria into the rectum of some of his patients with chronic gut infections, in order to straighten out their bacterial flora. He had some success with this procedure, and the results inspired others to try it as well.

In 1958, Dr. Ben Eiseman, Chief of Surgery at the Denver Veterans Administration Hospital, treated some very ill patients suffering from a diarrheal illness called pseudomembranous colitis, by giving them fecal enemas. This was very successful, even though it was not yet understood what exactly had caused the severe illness.

The patients had contracted an infection known as *Clostridium difficile*, a hard-to-say name for an even-harder-to-have bacterial infection of the colon that can cause inflammation resulting in abdominal pain, diarrhea and, in severe cases, toxic megacolon.

Clostridium difficile typically occurs in settings or events such as hospitalization, chronic illness, or recent antibiotic use—although it is becoming more common. It generally represents a shift in normal colon bacteria, allowing the *C. diff* (see, we got tired just

JUSTIN Toxic megacolon was always my favorite Teenage Mutant Ninja Turtles villain.

SYDNEE Toxic megacolon is a condition in which the colon becomes so inflamed that it basically stops functioning, allowing stool to accumulate in the colon as it gradually enlarges and distends the abdomen. This is a medical emergency and generally requires surgery to save the patient.

JUSTIN Well see, now I feel like a jerk.

typing the whole name) bacteria to overgrow the colon. The infection is only susceptible to a few antibiotics, and sometimes these don't even work.

Due to the severity of this infection and the possibly devastating consequences—as well as the fact that very few treatments exist—doctors and scientists have been searching for alternative ways to treat this infection for some time. The result has been increasing interest in fecal transplants as a way of introducing good bacteria back into the colon to help fight the bad, and the results have been very promising.

We could go on and on about medical uses for poop, much to Justin's delight, but we're assuming that you're currently distracted by the thought that every time you go Number Two you've been flushing away $40. We'll leave you to calculate just how much you're in the hole.

DO WE STILL DO THIS TODAY?

In the modern era, fecal microbiota transplant (FMT) is used for especially resistant cases of *C. diff* colon infection. The process is fairly simple, if less than appealing. Stool is taken from a donor (about 200 or 300 grams), who is usually someone known to the patient. It is then blended and introduced into the patient's intestine via an enema or nasogastric tube.

If you lack a willing donor known to the patient, you can pay one to give you a sample. Only certain donors are eligible to provide the fecal material, and they are very rigorously screened with stool and blood tests, as well as a general history and physical examination by a physician. Only certain consistencies of stool can be used as well. There is a chart for the classification of stool based on texture and density known as the "Bristol Stool Chart," and only samples rated a three to five are considered appropriate for transplant. All this will earn you $40 a sample, though, so it may be worth the effort. It is certainly less painful than selling plasma.

FMT has been used in some places for other causes of colitis with some success, and is being investigated as a treatment for inflammatory bowel diseases such as Crohn's and Ulcerative Colitis.

The FDA is still working on the process for regulating it, as it is treated more like a tissue than a drug currently. There are donor banks and regulations right now, but treatment is still considered investigational. Currently, businesses such as OpenBiome in Boston send poop all over the world to be used in treatment of *C. diff* infections.

THE WEIRD

When the going gets tough, the tough start coming up with the wildest solutions in recorded medical history.

A dance that can't stop

A room swinging with the sea

Cures bitter and sweet

The Dancing Plague

Shake it till you can't shake it no more—literally. That's a figure of speech; you can stop. Please? You're really scaring us.

Dancing can be a marvelous way to alleviate stress, a proto-mating ritual for awkward promgoers or—most nights of the week on Fox—a way to win fabulous cash and prizes. Dancing is one of humanity's most wonderful innovations . . . unless you lose the ability to stop.

You can't really blame Frau Troffea. After all, 1518 was a rough time to be alive, and Strasbourg (which was part of the Holy Roman Empire at the time, but is now in France) was certainly no exception. The region was wracked by famine, disease, and economic depression. Bread prices were the highest they had been in years. Syphilis had recently made an appearance, and leprosy and the plague were still hanging around.

Plus, it was July, and you know how muggy it can get in July.

So, we find it hard to fault Frau Troffea for stepping out into the street one (probably super muggy) July afternoon and beginning to dance.

It probably felt pretty good at first! Shake away those economic-and-plague-related, bread-starved blues, and get on down. We're assuming, however, it became a lot less relaxing the moment that Troffea realized she was unable to stop.

She continued dancing for four to six days, her feet bled, and she almost certainly attracted a crowd, but still she danced on. This would have been odd enough, but it was nothing compared to what came next: Other people started dancing as well. Within a week, with nary a pulsing rhythm or generous cash prize to urge them on, some thirty to forty people had joined the never-ending dance party. By the end of the month, 400 frantic bodies flailed wildly in the streets—and not a soul could tell you why it was happening.

This observation is probably in questionable taste, but is it fair to honor Frau Troffea as the inventor of the flash mob? And by "honor" I, of course, mean thoroughly and mercilessly castigate.

A TERRIBLE CASE OF BOOGIE FEVER

According to unnerving contemporaneous reports, no one seemed the least bit happy about this impromptu dance sesh. There were no smiles, no smoldering glances, and not the slightest bit of frolicking. For all their dancing, these people were clearly miserable. There are multiple accounts from the era of dancers begging for help even as they, apparently powerless to control themselves, pranced on.

The dancing continued day and night, rain or shine. They stopped for nothing.

City officials may have been content to let the dancing continue unabated for a while. Sure, there were a lot of people were missing work, but you couldn't find a ton of entertainment in those days, and watching a bunch of goofballs dance silently would have beat staring at the wall.

Except people started dying.

Well, I hope whoever's wedding proposal this flash mob was building to felt good about themselves. Are you happy now, Adolfus? Would it have killed you just to buy some flowers and get down on one knee? Maybe a nice dinner?

YOU MAKE ME FEEL LIKE DANCIN'

Soon enough, the exertion took its toll. Strokes, exhaustion, dehydration, and heart attacks began to fell the dancers, one by one. According to one report from the period, as many as fifteen people were dying every day while the dancing plague was at its apex.

City authorities knew they had to act. They consulted with priests, doctors, and anyone else that they could think of who might have some clue as to the cause.

The doctors ruled out the supernatural (always an important step that's so often neglected by today's physicians) and exonerated the moon from any wrongdoing. The best answer anybody could come up with? "Hot blood."

If you've read this far, your bloodletting sense is probably tingling. We're sorry to disappoint, but put away the leeches; no blood gets spilled

this time. Believe it or not, the great minds of Strasbourg were able to cook up a far, far stupider way of treating the dancing plague.

THE RHYTHM IS GONNA GET YA

The brilliant solution? Those afflicted needed to dance even more.

No, seriously, that was the prescription: more dancing. Physicians believed that if the dancers just danced enough, they would eventually dance out all their urges and impulses, and the dancing would finally cease.

So, Strasbourg's officials opened dance halls in public areas, and erected a stage in the town square. They even hired a band to provide music. Though that band had no official name at the time, we now know them as . . . The Beatles. And now you know the rest of the story.

No, no, we're kidding—The Beatles would never have played a gig that weird.

So, if you decided to go touring Strasbourg, would you still see a group of super old, super

JUSTIN: We scoff, of course, that's what we do, but do you have any better ideas for clearing out a perpetual dance floor?

SYDNEE: The usual?

JUSTIN: Well, sure, you can say that. But Phil Collins wouldn't be born for another 433 years, so no luck there.

tired people dancing to this day? Fortunately, no, though we've got no satisfying denouement for you. Eventually dancers who didn't die just stopped, one by one. The dancing plague ended as mysteriously as it had begun.

SO WHAT THE HECK HAPPENED?

The possible cause (or causes) of the plague are less murky today, but only slightly.

In 1952, author Eugene Backman decided to look for a chemical explanation for this phenomenon. He consulted with a few other experts of the day, and after some study, the likely culprit that they settled on was ergot (a type of fungus that grows on rye and similar plants). Ergot creates an alkaloid that causes a malady called "ergotism" in animals (like us) that consume grain contaminated by it.

Ergotism is typically associated with a highly unpleasant dry gangrene, but it can also cause convulsions, seizures, and psychosis. It's been theorized that ergotism was mistaken for bewitchment in the 1690s, disastrously prompting the Salem witch trials.

In his book, *A Time to Dance, A Time to Die: The Extraordinary Story of the Dancing Plague of 1518*, author John Waller contests laying blame on ergotism, pointing to the fact that the disorder's spasms wouldn't be mistaken for coordinated, dance-like movements that went on for days.

Sociologist Robert Bartholomew theorized the dances could have been some sort of celebration or demonstration by a pagan or heretical sect. Waller also refuted the suggestion, because that would suggest the dancing was intentional, and by all accounts, every dancer was desperate to stop.

Really? Heretical dancing? So we've now moved from "vaguely reminiscent of *Footloose*" to full-on copyright infringement territory.

Okay, Waller, Mr. Smart-Face, if you know so much about dancing plagues, what's your theory? Oh. Actually, he has a pretty good one. Waller believes that dancing plagues (yes, there have been multiple occurrences throughout history) are caused by a sort of collective freakout known as "mass psychogenic illness."

Things, as we've noted, were really crappy at the time, and fear and anxiety were high, which is prime breeding ground for mass psychogenic illness, a phenomenon often preluded by huge psychological distress.

There was a legend at the time, a Christian martyr named Saint Vitus could curse people who displeased him by forcing them to dance. So while the dance was not caused by supernatural forces, the fear of being cursed to dance could have triggered the response. Also, as a random side note, St. Vitus is considered to be the patron saint of dancers and entertainers. Go figure.

Since this is a disease that spreads psychically, putting a bunch of dancing people on display in the middle of town is basically the worst thing that the officials could have done. That's right: It's an even dumber idea than you initially thought a few paragraphs ago.

IS THIS STILL A PROBLEM TODAY?

Perhaps the best indication that the "dancing plague" was a mass psychogenic illness is that it was basically a fad, though a centuries-long one. As science began to explain more of how the world worked into the 1600s, fear of things such as curses began to decrease. Without the prevalent fear of being cursed to dance, the "curse" didn't manifest.

Thank goodness we live in this new, bold, educated age when our young people are too busy attempting to eat detergent pods or handfuls of cinnamon to mindlessly dance in the street. Hooray for enlightenment.

DANCING IN THE STREETS

The Strasbourg dancing-plague outbreak was perhaps the largest in recorded history, but it was by no means the only one. Here are a few other notable cases.

RIVER MEUSE, GERMANY (*1278*) Two hundred people started dancing near the River Meuse in Germany. They eventually ended up on a bridge that apparently wasn't rated for 200 hoofers, because it collapsed, killing several of the dancers.

BERNBERG, GERMANY (*1020*) Eighteen peasants began dancing around a Christmas Eve service at church and disturbed it. (Yes, we're as surprised as you are that the harrowing tale of eighteen people doing unexplained Christmas dancing has persisted for nearly a millennium.)

ERFURT, GERMANY (*1237*) A group of unruly kids started dancing and boogied their way from Erfurt to Arnstadt, Germany. Google says that's a four-hour-and-two-minute walk; who knows just how long a Macarena that was?

THE BEAT GOES ON

Small towns all over Europe had isolated dancing outbreaks. Germany, France, Luxembourg, Italy, and Holland all were affected. A monk died in Schaffhausen, and a bunch of women entered a dancing frenzy in Zurich. This continued, scattered, for decades.

Not all these plagues were created equal. They all tended to occur during times of great shared stress, but some of the particulars varied. In some cases, dancers got naked; in others, they dressed in bright colors or put garlands in their hair. In yet other cases, they screamed while others sang; some had sex; some laughed. Some would become violent with those who wouldn't join in. In some cases, people would travel long distances to join in, and in others, it was just one guy dancing. Dancing plague victim or just a weirdo? You be the judge!

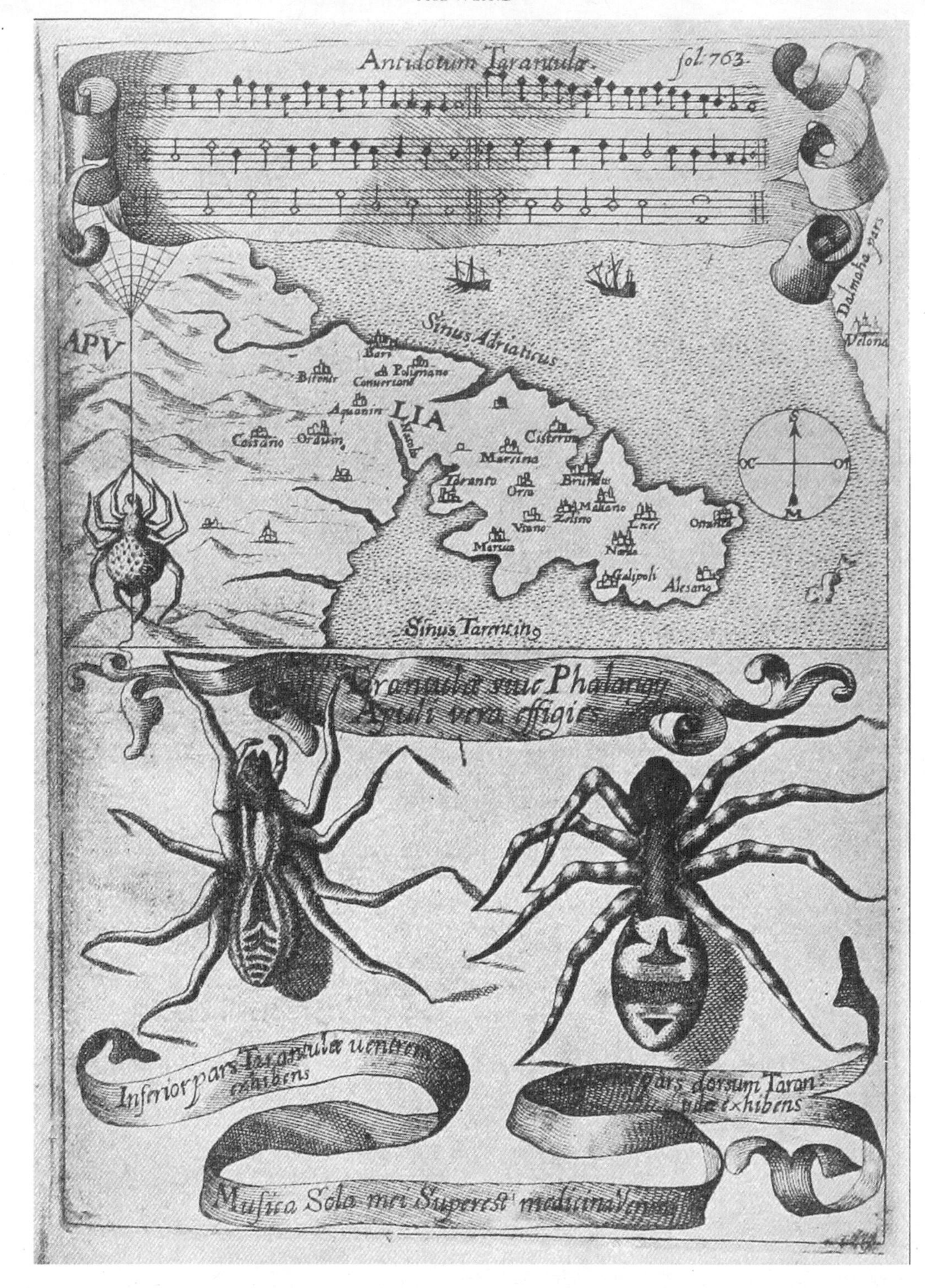
Antidotum Tarantulæ.
fol. 763.
Dalmatia pars
Velona
Sinus Adriaticus
APV
LIA
Bari
Taranto
Galipoli
Alesano
Sinus Tarentin9
Apuli vera effigies
Inferior pars Tarantulæ ventrem exhibens
pars dorsum Taran-
tulæ exhibens
Musica Sola mei Superest medicina

AN INFECTIOUS SENSE OF FUN! (EXCEPT, NOT FUN AT ALL)

Boogie fever isn't the only kind of psychic disorder to have appeared and then vanished. Here are some other notably odd behaviors that turned inexplicably contagious.

Tarantism In southern Italy from the 13th to the 16th century, it was believed that victims of a certain legendary spider called a "tarantola" could only be cured by dance. Yes, this is another dancing disorder, but this time shaking a tail feather is the cure, not the disorder. Other cures included drinking large amounts of wine, jumping into the sea, being tied up and whipped with vines, and fake sword fighting. Fun fact: There's a theory that tarantism led to the creation of a type of upbeat folk dance called the Tarantella, but there's a competing theory that the dance was actually the creation of an ancient Bacchanalian cult driven underground by the Roman Senate. Tarantism, according to the theory, was a fable created to mask the dance's resurgence.

Tanganyika Laughter Epidemic In 1962, a mission school in what is now Tanzania had to be closed on account of laughter. It started when three girls started laughing, and pretty soon almost the entire school was cracking up. The Chuckle Plague, a catchy name that we just now coined, lasted over two weeks, shut the school down, and even spread to nearby villages and schools. By the time it had run its course, fourteen schools and a thousand people were affected over six months.

Koro The belief that one's genitals (or, in rarer cases, nipples) are shrinking is one that has emerged sporadically in Asia, Africa, the United States, and Europe. The flare-ups are typically tied to widespread anxiety or social tension, and also appear to be exacerbated by media coverage or public attention. That said, there is some evidence that koro epidemics aren't necessarily connected to one another. That is to say, this very specific paranoia has emerged independently in several different cultures that have had no prior knowledge of the previous outbreaks elsewhere.

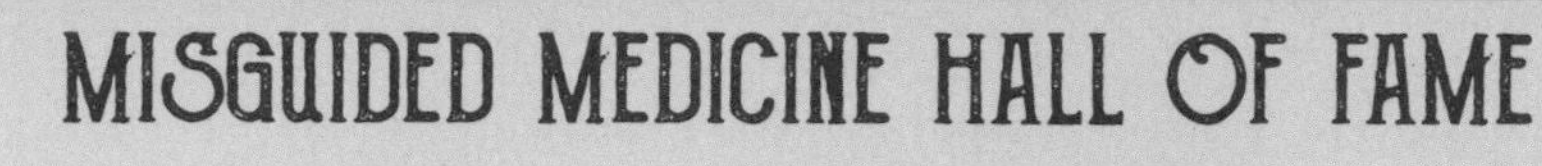

MISGUIDED MEDICINE HALL OF FAME

CURTIS HOWE SPRINGER

1896–1985 · USA

As we've seen, hucksters will go to great lengths to con unsuspecting rubes into spending heaps of money that they don't have on ineffective medical treatment. But who among them would be willing to found their own town to sell their wares? Only Curtis Howe Springer, the King of Quacks. (That's the actual title bestowed upon him by the American Medical Association, by the way. They . . . weren't fans.)

Springer's life was so buckwild that we're going to have to give you the abbreviated version:

Two years of high school. One year of bible school. Taught boxing. Sold sheet music. In the 1930s, he started giving lectures claiming to be either a representative from The Springer School of Humanism, the National Academy, the American College of Doctors (all fake) or the dean of Greer College (real, but closed), and started calling himself a minister. By 1934, he applied for airtime in Chicago so he could start selling fake medicine, and sing gospel songs.

Howe's fake meds included the Hollywood Pep Cocktail, an antacid called "Re-Hib" (which was mostly baking soda), "Delicious Manna" (a vegetarian food supplement), and a DIY home hemorrhoid treatment.

He really started spreading his wings with a number of health spas under the name "Haven of Rest" in Pennsylvania, and later in Maryland and Iowa. They all closed rather quickly, typically for failure to pay taxes. Howe needed his own haven of rest from Uncle Sam's sticky fingers, so he filed a mining claim on some land in the Mojave Desert and began creating his dream spa. He called the town . . . Zzyzx (so that it would be "the last word in health care").

Luxuriate in the beautiful concrete structures that were built by homeless people Howe hired from Los Angeles, so he could finagle his way into tax-exempt status (despite the fact that he only paid them in room and board).

Soak in the astounding "hot spring"! Did we put legally-mandated quotes there because it was just a boiler under some pools? We'll never tell! (But yes, that is the case.)

Witness some of the beautiful landmarks including a food-processing plant, a church, a man-made lake, an airstrip, and a sixty-room hotel called "The Castle" that Howe built for himself (because of course he did) on the road named "The Boulevard of Dreams" (because of course it was).

Rejuvenate in our spa where you can indulge in really expensive laxatives and feast on rabbit and homemade ice cream. Later, slather a mineral salt paste Howe called "Zypac" on your scalp, bend over, and hold your breath! When you feel flushed, you'll know it's working.

Listen to our on-site radio station broadcasting all the latest and greatest in gospel music and fake medicine ads—and requests for donations.

Refrain from alcohol and arguing. Sorry, those are the rules there.

Stay in one of the lots we're happy to sell to visitors so they can become permanent residents and live right near the spa. Is that legal to do on a mining claim? We're sure the Feds won't mind!

Leave in the late 1960s when the Feds realize they very much do mind and decide to see how much "mining" Howe had done on his mining claim. (It was none, obviously—unless he found some ore when he was digging the lake.) The U.S. government stripped him of everything and converted Zzyzx into the Desert Studies Center, a research outpost that's a lot less fun to visit, but a heck of a lot easier to spell.

Smoke 'Em if You Got 'Em

Everybody knows that smoking, dipping, or even hookahing tobacco is a terrible idea if you, like most of us, are allotted only one human body for use on Earth. But what if, this whole time, we've just been ignoring the incredible health benefits of this unfairly maligned and highly addictive plant? Wouldn't that be an amazing discovery?

It might be, but that's irrelevant, because we haven't been ignoring any such benefits, no matter what folks may have thought at one time. But hey, we've already started this section, so we might as well just keep on trucking.

The Innovators

When Europeans first encountered Indigenous Americans burning tobacco, it was already being used for both pleasure and (medical) business. Early on, native people realized that inhaling too much of the burning plant's smoke would feel great right until it knocked you out, a side effect that may have been used as an anaesthetic for trepanation. It was also likely used as a toothpaste after being mixed with lime and salt. (Tobacco toothpaste, called "creamy snuff," was actually sold in India until it was outlawed in 1997.)

The Treatment

After explorers brought tobacco to Europe, it was used for a staggering number of maladies largely based on real (and fabricated) accounts of its use by natives in the "New World." Tobacco was mixed into tonics designed to prevent hunger and thirst, to treat colds and fevers, to aid in digestion, and to heal skin lesions. In 1747, John Wesley's Primitive Physick prescribed tobacco smoke for earaches—yup, just blow it right in the ear—and for hemorrhoids (no, we don't know if you blow it in there as well. Stop asking.)

One of the problems with tobacco being used as a medicine, besides the whole being poisonous and giving you cancer thing, is that it's a stimulant. That means that tobacco products actually do have an obvious effect when ingested. In an era when so many "treatments" did absolutely nothing, it was easy to buy its efficacy for all manner of illnesses based on that quality alone.

How'd That Work Out?

As early as 1602, doctors were raising concerns about using tobacco. It was addictive and harmful if abused, and it was being prescribed higgledy-piggledy, with very little concern for proper dosing, let alone actual efficacy. By the late 1800s, as the real dangers of tobacco became evident, medical use largely died out.

Tobacco doesn't play much of a role in modern medicine, unless you happen to count "dastardly villain." There has been some research into a possible connection between cigarette smoking and a reduced risk of Parkinson's, but the risks still vastly outweigh whatever potential benefits there may be.

FUN FACT: In 1560, the French ambassador to Lisbon, Jean Nicot, was gifted a tobacco plant by a warden during a prison visit. After one of his pages used the leaves to treat a skin lesion, Nicot started hailing the plant as a miracle cure. He foisted it on so many people, it was eventually renamed in his honor: Nicotane. We spell it with an "i" now, and use it to refer to tobacco's active ingredient, but the name has stuck around to this day.

A Titanic Case of Nausea

If this saloon's a'rockin', don't come a'knockin'—it's made to relieve sea sickness via a system of gimbals.

While it's true that you'll find more ill-advised, more life-threatening, more disgusting and, yes, more plain old stupid treatments in the annals of medical history. But we're fairly certain you won't find another one that caused this much property damage.

"Well, you did it, Crom. I know I doubted you along the way, but here we are, standing on the water and not falling. This creation of yours is astounding. What did you call it again? A boat?"

"Yes, my friend, it's called a 'boat.' Now that we're out here though, I—" [barfs uncontrollably]

Okay, maybe the relationship between boats and seasickness doesn't go all the way back to caveman days. But we're betting it wasn't too long after the first bottle of champagne was smashed across a ship's hull that the first bottle of Dramamine was less dramatically uncorked.

But, of course, we didn't have Dramamine in the early days of maritime adventure (we didn't have champagne either, but that's some other book's problem). This is the story of our ill-fated attempts to cure seasickness, and how they led to one of history's dumbest boats.

BETRAYAL ON THE HORIZON

So, what is it that causes you to feel a little wobbly out on the waves?

Well, it's simply confusion between your visual input, and vestibular and position sensors in your inner ear. Easy! Don't worry—we only half understood it ourselves (the Sydnee half). Let's break it down: You see a horizon moving one way, but the rest of your body feels itself moving up and down independent of that. Additionally, your legs are adjusting to keep you from falling, and well, it just gets to be too much, especially when parts of the ship aren't moving.

Though not everybody gets desperately ill on the waves, it's rare to find someone who's completely unmoved by this potent cocktail of stimuli. In fact, only 10 percent of people are completely impervious to motion sickness of some sort. (And we would bet that a few of them are just lying to look cool.)

HEAVING THROUGH HISTORY

When tracking a disorder all the way back to the beginning of history, we often like to duck in on the ancient Greeks to see what they have to offer. It's fun to see all the flowery ways they find of making crap up. The word nausea comes from the Greek *naus* for "ship"—certainly Hippocrates has something spicy to say on the topic, right? "Sailing on the sea," he wrote, "proves that motion disturbs the body."

Uhh, good job, Hippocrates. Try to rest up, pal; we'll try to catch up with you next chapter.

A few hundred years after Hippocrates stated the obvious without offering any kind of cure to go with it, Roman orator and politician Marcus Tulluis Cicero said he would rather be killed in battle than deal with seasickness.

While the Romans and Greeks groused, ancient China and India had both found that ginger offered some relief to seasickness. A study in the *Lancet* has shown that powdered ginger staved off nausea twice as long as a placebo.

And that's the end of the chapter.

No, no, we still have a lot of dumb historical stuff to get out of our systems and into your brain.

A WORLD-CHANGING TUMMY ACHE

Much is made of the Black Plague and its power to shift the course of human history. But humble seasickness deserves at least a footnote for the role it played in helping the English defeat the Spanish Armada in 1588.

The Spanish Admiral Alonso Pérez de Guzmán was more of an administrator than a sailor, and hadn't spent much time at sea. He wrote to the king of Spain during the campaign, "I know by the small experience I have had afloat that I soon

It's bad enough when poor schlubs like us suffer from a medical condition, but famous historical figures? Seasickness had gone too far. It was time for the medical establishment to fight back the only way it knows how: badly, followed by excruciatingly slow improvement.

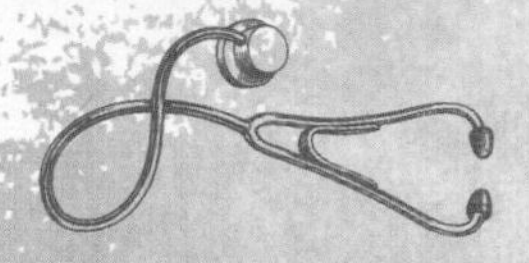

Sydnee's Fun Medical Fact

If you want to sound really fancy, you could always call it a *mal de mer*. That also means seasickness, of course, but it will probably seem much more impressive to your friends when you start sharing all your interesting new tidbits about nautical nausea.

become seasick." Reportedly, his *mal de mer* was one of the factors that led to his loss at the hands of the British Navy.

Guzmán wasn't the only chronicler of his own seasickness. Charles Darwin, Lawrence of Arabia, and hero of the British Navy Admiral Horatio Nelson all journaled their struggles with tummy troubles on the water. That's surprising, but maybe not so surprising considering one old English proverb states that "The only cure for seasickness is to sit on the shady side of an old brick church in the country." Poor Nelson never even stood a chance.

THE BATTLE AGAINST BARF

Bad treatments almost always start from bad ideas about how the human body works, and those early docs were working with a doozy of one. Namely, the prevailing belief through most of history was that shifting intestines and stomach caused the symptoms. So, the treatments for seasickness were largely concerned with keeping all of your internal organs in place.

In the 19th century, a Canadian company made an anti-motion sickness belt, a tight, girdle-like number that held all the gut plumbing in place. It was advised that those especially prone to seasickness could try wiring them to batteries so they could shock themselves if any nausea began, an innovation anyone who's read this far in the book shouldn't find too surprising. (Also, not too surprisingly, the belt didn't work.)

Even once we gave up on trying to keep our organs under control, solutions weren't much better. The Hamburg American Steamship Company took a pass at the problem with a vibrating anti-seasickness deck chair. How did that work? According to a 1906 edition of New Zealand's *Poverty Bay Herald*: "It is claimed for the chair that the up and down vibratory movement renders the pitching, heaving, and rolling of a vessel less perceptible, as the lengthy downward motions of the ship become neutralised by the rapid succession of vibrations imparted from the chair."

So . . . it didn't work. That's the answer.

If you asked Henry Bessemer, the Hamburg American Steamship Company was simply thinking too small. You couldn't make a simple deck chair seasickness-proof.

It had to be a whole ship.

THINKING BIG REALLY BIG

We're going to talk about this buckwild invention in a second, but let's take a moment to meet your new pal Henry Bessemer.

The hero of this tale was born in England in 1813 and followed in his father's footsteps as a wealthy inventor.

He didn't invent the same stuff again though, right? That'd be pretty unimpressive.

Bessemer's most notable contribution to society was the cleverly named "Bessemer method," which made steel inexpensive to produce for the first time in history.

SIR HENRY BESSEMER IN HIS 80TH YEAR

An important fella like that was in demand all over Europe, which meant Bessemer spent a lot of time traveling back and forth across the English Channel. He was sick on every trip. Like the hero of every great infomercial, he thought aloud with frustration to no one in particular: "There has to be a better way!"

Bessemer got his brilliant anti-nausea idea from watching a compass, which stays still despite the pitching and rolling of the ship. He thought, logically—if perhaps fantastically—if we can make a compass that doesn't move, why not a whole room he could hang out in on the ship?

He built a model in his backyard in London, which must have been quite a treat for the neighbors. The cabin was supported by gimbals—a mechanism, typically consisting of rings pivoted at right angles for keeping a compass or chronometer horizontal in a moving vessel, and not attached to outside walls of the ship. Basically, the saloon could move independently of the ship so it stayed level.

While the term "saloon" is typically just a way of saying "old-timey bar," we're using it in the naval sense here, specifically "a large cabin for the common use of passengers on a passenger vessel." That's not to say alcoholic drinks weren't consumed in the Bessemer saloon. You're on a boat, what the heck else are you going to do?

He did some tests to see if this would work in his backyard (how exactly the hell he did that has tragically been lost to history) and he was pleased. So, he had one of his cabins installed in

a steamship with the assistance of a ship designer. It was seventy feet in length, thirty feet wide, and was extremely fancy. It had gold-plated mirrors on the walls, leather seats, and potted plants everywhere. Very Victorian chic.

The fact that it's extremely fancy is gonna make this next bit extra fun to imagine, so make sure you've developed a clear mental picture before we move on. Or just look at these actual pictures from newspapers of the time.

THE SEASICK-PROOF SALOON SETS SAIL

Join us now for the maiden voyage of the *SS Bessemer.* The year is 1875, and the ship is packed to the portholes with a super high-class, privately invited guest list of investors, and assorted other rich people.

They departed from Dover, and headed for France, planning to land at the Port of Calais. Everything was going pretty well and the saloon was miraculously stable. It seemed as though Bessemer had done it!

But then the ship began to slow down as it entered the harbor.

Remember how the saloon was swiveling independently of the ship itself? Well, it didn't get the memo that it should cut it out when the ship tried to stop moving. The swinging of the cabin as the ship slowed down made it incredibly difficult to pilot the ship. And lo, did the *SS Bessemer* end its maiden voyage by crashing into the pier.

For a lesser inventor, that'd be the end of the story, but not so for Henry Bessemer. He rushed back to England, and after taking just a month for repairs (and so he could drum up replacements for

the investors who had bailed), the *SS Bessemer* set sail on its first public voyage.

The ship had one important difference this time: The saloon was locked in place.

Okay, so it kind of defeated the purpose to lock the saloon in place, but I get why he'd be desperate to get a win on the books at this point. You know when you're baking banana bread, and then you burn it, but then you eat it anyway, because hey, at least you're full? This was the burnt banana bread lunch of sea voyages.

Bessemer claimed that it was locked like this because he had insufficient time to fix the damage after the first trip. Sure, Hank. Let's go with that explanation.

The *Bessemer* fared pretty well once again, but the real test wouldn't come until the ship slowed as it prepared to dock. Crowds gathered at the Port of Calais to see the *Bessemer*'s triumphant arrival or disastrous slow-motion crash. Considering the crowds were comprised of human beings, we assume they would have been equally thrilled with either outcome.

They got the second one.

As the *Bessemer* slowed, it began to move erratically, locked saloon be damned. The captain fought to keep control of the ship, but it once again crashed into the port—this time taking out most of the supporting pillars and destabilizing the entire pier.

This second crash was understandably enough to scare off the investors who remained, and the ship was left to rust in port. Bessemer finally abandoned the idea altogether.

The dream lived on though, albeit in a completely stationary sort of way. Years later, when the Bessemer was broken up by British naval architect Edward James Reed, he had the saloon moved to his home in Kent to use as a billiard room. His house would later become the Swanley Agricultural College, and the saloon was used as a lecture hall.

Tragically, the saloon met its final end, not at sea, but after being destroyed by German bombers during World War II. This is probably not even on the top 1,000 list of bad stuff Nazis did (even if you ignore them trying to steal the Ark of the Covenant), but still, the destruction of such a lavishly appointed construction was, in Victorian terminology, a real bummer.

DO WE STILL DO THIS TODAY?

What, build special anti-sea-sickness boats? No, what a silly question. We've made strides against the malady in other ways, though! By observing that people with hearing deficits were less likely to become seasick, we began to understand that *mal de mer* had some connection to the inner ear.

The Second World War increased efforts to find something that helped—the least it could do after claiming the Bessemer saloon, really—and in 1947, we discovered that antihistamines could help prevent motion sickness. This was the breakthrough humanity had been waiting for.

These days we use antihistamines, such as Dramamine and Phenergan, to help treat seasickness. Again, ginger helps too, if you want to go the natural route.

You can also change your boat behavior to help limit the effects. Don't eat a lot or drink booze, watch the horizon, and try to stay below decks and near to the middle of the ship you're aboard.

But for heaven's sake, if you do decide to make a boat to treat your seasickness, try to limit your ports of call to only the most resilient of piers.

Years of reading way too many detective stories have left us with what we describe to friends and family as "an altogether healthy and reasonable distrust of arsenic." But when one doctor looked at the poison of literary legend, he saw a lifesaver . . . and he might have actually been on to something.

The Innovators

In 1786, English physician and inventor Thomas Fowler identified the secret ingredient in a popular line of "malaria drops," patented by chemist Thomas Wilson five years earlier. It was arsenic, but you've almost certainly guessed that, unless you're the sort of person to skip past chapter titles—and that just doesn't sound like you! Fowler decided to put his own spin on the potential poison.

The Treatment

Fowler created a solution of one percent potassium arsenite, which he called "liquor mineralis," for "agues, remittent fevers, and periodical headaches." ("Agues" is an old-timey description of a disease that includes fever, but usually refers to malaria. Wilson's drops were actually designed as a treatment for agues.) In 1809, liquor mineralis, known by that time as "Fowler's Solution," was accepted into the London Pharmacopeia and became widely used as an alternative to quinine for malaria and as a cure for "sleeping sickness" (trypanosomiasis). By the 1880s, Fowler's Solution was used for a variety of other ailments, including asthma, eczema, psoriasis, anaemia, hypertension, gastric ulcers, heartburn, rheumatism, and tuberculosis.

We're understandably pretty suspicious of panaceas here at the Sawbones Lodge, and Fowler's Solution is no exception. But this one comes with an interesting coda.

While doubtless Fowler's Solution wasn't effective in treating all the stuff it was supposedly good for, in 1878, the concoction was discovered to lower the white cell count in chronic myelogenous leukemia. In layman's terms: It worked on leukemia. It worked so well, in fact, it was the main treatment for it until the advent of radiation and chemotherapy in the 20th century.

How'd That Work Out?

Throughout the first half of the 1900s, we became progressively more distrustful of cure-alls, and they gradually fell out of favor as we began to understand how certain medicines targeted specific problems. That fall from grace went double for cure-alls like arsenic-based medicines that were … well, poison.

But there's some evidence that medical history may not quite be done with Fowler's formula. In the past couple of decades, Chinese scientists have been experimenting with arsenic as a treatment for relapsed acute promyelocytic leukemia (APL). By September 2000, the U.S. FDA had approved the arsenic-based medicine known as Trisenox for use. China saw an encouraging nine out of ten patients responded to the treatment.

FUN FACT: When you're thinking about arsenic's poisonous properties, you're thinking of an oxide of the element called, appropriately, arsenic trioxide. It's formed when oxygen comes in contact with arsenic, and (here comes the fun part) it gives off a garlic-like odor! Well . . . we thought it was fun.

No, wait, we've got something: What if vampires aren't actually allergic to garlic, they're just super paranoid about arsenic? See, fun right? Yeah, we thought so, too.

MISGUIDED MEDICINE HALL OF FAME

PARACELSUS

1493–1541 · SWITZERLAND

This Swiss German physician wandered the world (often because he'd been ostracized from a university or entire town) trying to find colleges that were worth his time—and found them all wanting. He once famously said he didn't know how "the high colleges managed to produce so many high asses." Ugh. If Bob Dylan and vinyl had existed in the time of Paracelsus, he would have chastised you for listening to *Highway 61 Revisited* on anything else.

His full name before he renamed himself with a dunk on poor dead Celsus? Philippus Aureolus Theophrastus Bombastus von Hohenheim. Try saying that three times fast, huh?

Paracelsus wrote one major work, a book about surgery. Guess what he called it? *Die Grosse Wundartznei* which actually, no kidding, means *The Great Surgery Book*. Literally.

You wanna know the most irritating about Paracelsus? Every once in a great while, he came up with an idea that shook the foundations of medicine. The rest of the time, he was just being a total jagweed. We've collected some of Paracelsus' most notable beliefs and creations for you here.

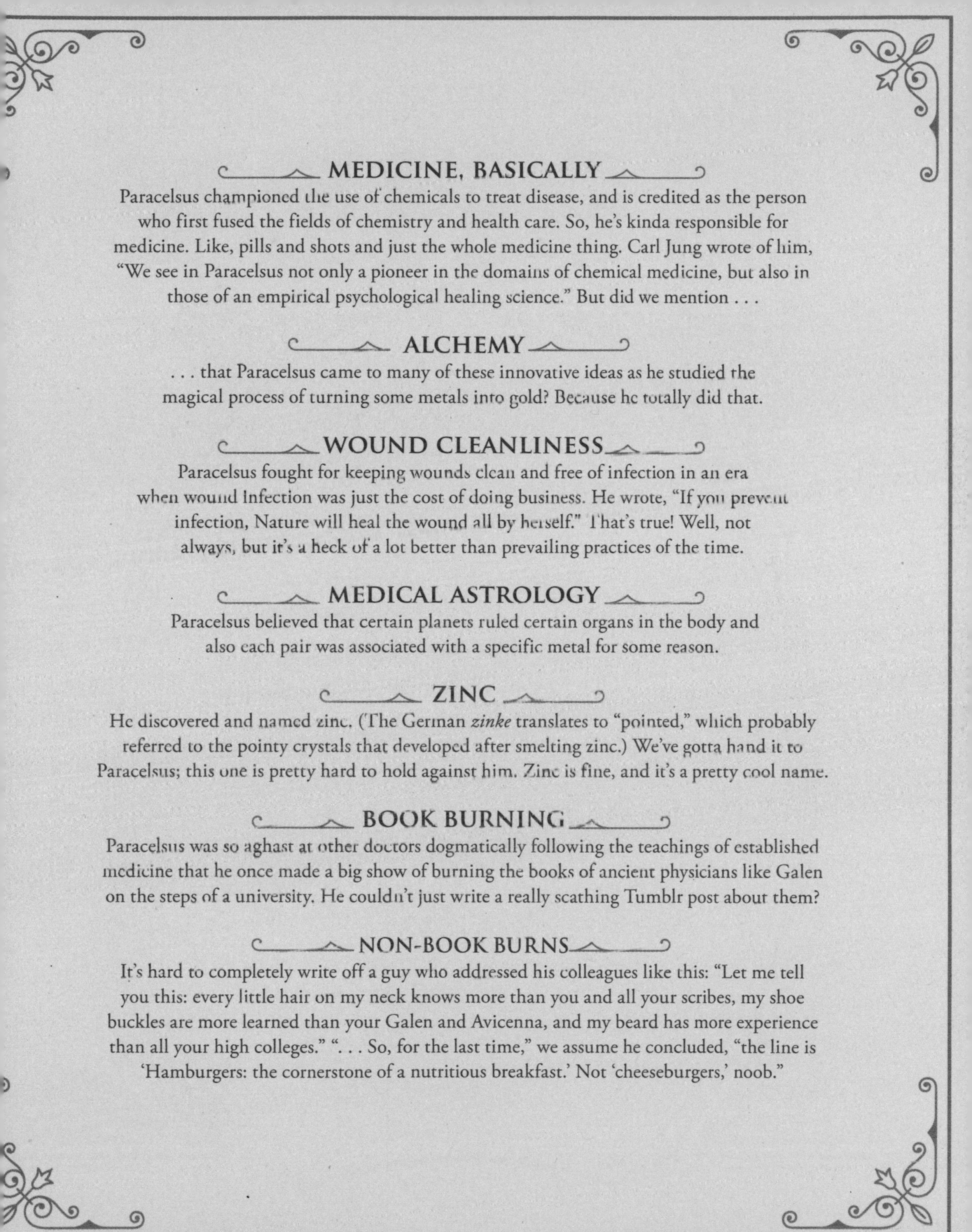

MEDICINE, BASICALLY

Paracelsus championed the use of chemicals to treat disease, and is credited as the person who first fused the fields of chemistry and health care. So, he's kinda responsible for medicine. Like, pills and shots and just the whole medicine thing. Carl Jung wrote of him, "We see in Paracelsus not only a pioneer in the domains of chemical medicine, but also in those of an empirical psychological healing science." But did we mention . . .

ALCHEMY

. . . that Paracelsus came to many of these innovative ideas as he studied the magical process of turning some metals into gold? Because he totally did that.

WOUND CLEANLINESS

Paracelsus fought for keeping wounds clean and free of infection in an era when wound infection was just the cost of doing business. He wrote, "If you prevent infection, Nature will heal the wound all by herself." That's true! Well, not always, but it's a heck of a lot better than prevailing practices of the time.

MEDICAL ASTROLOGY

Paracelsus believed that certain planets ruled certain organs in the body and also each pair was associated with a specific metal for some reason.

ZINC

He discovered and named zinc. (The German *zinke* translates to "pointed," which probably referred to the pointy crystals that developed after smelting zinc.) We've gotta hand it to Paracelsus; this one is pretty hard to hold against him. Zinc is fine, and it's a pretty cool name.

BOOK BURNING

Paracelsus was so aghast at other doctors dogmatically following the teachings of established medicine that he once made a big show of burning the books of ancient physicians like Galen on the steps of a university. He couldn't just write a really scathing Tumblr post about them?

NON-BOOK BURNS

It's hard to completely write off a guy who addressed his colleagues like this: "Let me tell you this: every little hair on my neck knows more than you and all your scribes, my shoe buckles are more learned than your Galen and Avicenna, and my beard has more experience than all your high colleges." ". . . So, for the last time," we assume he concluded, "the line is 'Hamburgers: the cornerstone of a nutritious breakfast.' Not 'cheeseburgers,' noob."

When your oatmeal is a little bland, or you realize that drinking unsweetened tea is pointless, or you want something to dip your chicken nuggets in that will really trip people out but is surprisingly delicious, what do you reach for? What? No, not mayonnaise. Why would you say ma—ugh, never mind.

We're talking about honey.

But before you reach for that honey, consider: Shouldn't you save it for a real emergency? You know, like sudden-onset diabetes, or baldness, or any of the countless other diseases honey has been prescribed for over the years.

So what, you've probably never asked yourself, exactly is honey? Well, friend, it's bee vomit. You've (probably) eaten bee vomit. We don't have a cute follow-up; we thought we'd let you process at your own pace.

Anyway, charming as this intro has been, we've got too many totally real, legitimate uses for honey to share to waste anymore of your time.

USED TO TREAT:

IMMORTALITY

Okay, so that's an overstatement. But ancient Greeks like Aristotle really did believe that consuming honey could extend your lifespan. The health spa of the era, the *Asklepieion*, even offered a honey therapy.

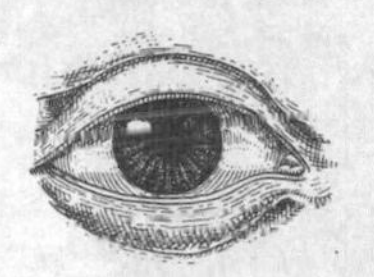

CATARACTS

Mayan priests have used honey to treat cataracts for hundreds of years. Fun fact: Mayan healers get their honey from a stingless bee cultivated and worshipped by ancient Mayans. That doesn't mean the honey will work, but hey, at least you won't get stung in the process!

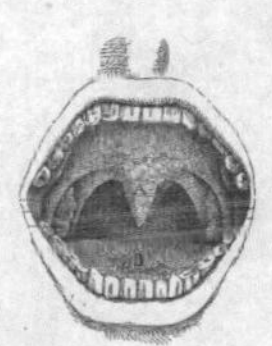

COUGH

Ancient Ayurvedic text listed honey as a treatment for cough (among many other disorders). This one isn't off the mark: Honey has demulcent properties, which is a fancy way of saying that it creates a film on mucus membranes, which can be soothing to your throat, if only briefly.

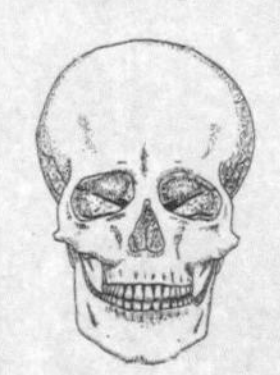

BALDNESS

Oops, nope, you choked. Honey mixed with cinnamon is a popular folk remedy for hair growth. It's a quick path to a sticky head and not much else. The only way this would work is if the honey is being used as an adhesive for actual hair.

CONSTIPATION

In *Compendium of Materia Medica*, a foundational Traditional Chinese Medicine text by Ming Dynasty pharmacist Li Shizhen recommends eating a daily dollop of honey to relieve and prevent constipation. Honey has been proven to have a mild laxative effect in some people. Ancient medicine, you're two for two! This your best hot streak of the book— so don't screw it up!

WOUND HEALING

It's rare to find an ancient medical tradition that doesn't include honey for wound care. And surprisingly (but not so surprisingly; considering honey's weird hit rate this chapter), it's not a bad choice. It makes a barrier for the wound, it promotes drainage through osmosis, it's antiseptic, it prevents dressings from sticking to wound surface, it's anti-inflammatory, and it even makes the wound smell better. It's great!

In fact, the only reason we stopped using honey as a wound treatment so darn much is that we discovered antibiotics, and got a little cocky. But now, as the effectiveness of antibiotics is starting to decrease in some cases, we're starting to investigate honey's strengths as a wound treatment once again. That's both a cool bit of trivia about honey and also totally terrifying!

(By the way, those properties are ascribed to medical-grade honey, which is a thing. Don't cut yourself and go hunting for the plastic bear to squeeze on it. It won't turn out well.)

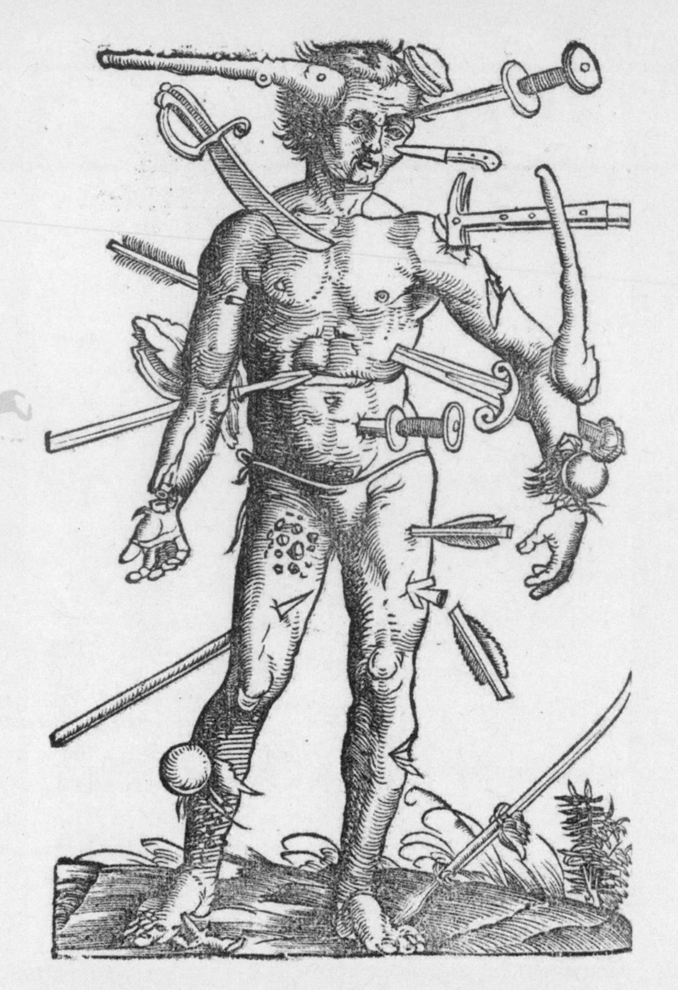

Self Experimentation

Physician,
heal
thyself!

. . . or, you know, give it a shot by injecting yourself with all manner of weird substances and tinctures, and then just hope against hope it doesn't kill you. And even if it does, maybe you'll end up in a book of zany medical history ephemera. Win win!

You thought you could rely on your friends and family. You carefully explained how close you were to unlocking a miraculous cure to some dread disease. All that you needed to do now was to inject them with your homebrew concoction of bee pollen, moonshine, and chewed-up bubble gum. But did they roll up their sleeves like good little test subjects? No, they did not.

Then you pinned your hopes on starving college students. You gave them the exact same pitch about saving lives, but sweetened the pot with a Target gift card worth $35. Still nothing.

Looks like you're left with no other choice, friend: You're going to have to experiment on yourself. At least you're in good company, which is to say the heroes who did something brave (and maybe just a little stupid) to help advance the human race's battle against disease.

STARTING OFF WITH A BANG

London, 1787

Self-experimentation, understandably, has had a tendency to break bad for a lot of practitioners. That said, it's hard to imagine it breaking much badder than it did for mid-18th-century physician John Hunter.

Originally from Scotland, Hunter moved to London at the age of twenty-one, in 1749. The city had been largely rebuilt after the devastating Great Fire of 1666, becoming a bustling early modern metropolis with one of the world's busiest seaports. All these factors no doubt contributed to a rising number of prostitutes in the streets and, perhaps related, a growing number of patients presenting with symptoms of venereal disease.

Most physicians of the era believed, correctly, that they were seeing two distinct diseases in their unfortunate patients: gonorrhea and syphilis. Hunter wasn't buying it. He was certain that all of the symptoms were caused by a single infection by a single pathogen.

He theorized that gonorrhea was passed from person to person via what he called a "venereal poison." It then, according to this hypothesis, spread throughout the body and just sort of became syphilis at some point.

Hunter determined that he had to test this theory on someone who'd never had either disease, so he picked . . . himself.

LAST CHANCE TO BAIL:

Oh boy, y'all, this one is tough! Whether you've got a penis or you don't have a penis, pretty much any penis-status person is going to be equally unenthused about this.

Hunter made cuts in his own penis, and then injected pus from a gonorrhea patient's penis into the cuts made in his own aforementioned penis. . . . Are you okay? Do you need to breathe into a paper bag or anything?

So, Hunter was "successful" in giving himself gonorrhea, which he apparently counted as a win. And then things got even better (for his theory if not for his junk) when he developed a chancre, a specific kind of sore that's frequently a sign of syphilis. And sure enough, it was. This convinced the good doctor that his incredible sacrifice was justified, in the name of proving himself right and advancing medical science. He even got to name it the "hunterian chancre."

Here's the twist: He was wrong. The patient he took the samples from just happened to have both diseases already. So, Hunter's widely disseminated findings actually ended up setting the study of venereal disease back for some time.

So, to recap: John Hunter cut up his wiener, blasted it with gonorrhea/syphilis pus, and all in the service of . . . setting back medical progress.

Oh, right, he also got gonorrhea and syphilis. Cool Thursday, John!

NOT TOTALLY GROSSED OUT YET? READ ON

Philadelphia, 1790s

If you suspect that a "trainee doctor" in the 1700s with a name like Stubbins Ffirth, would be just plain N-A-S-T-Y, you'd be right. Just how nasty do you have to be to demand that we spell out "nasty" in all caps? Read on.

In 1793, a massive yellow fever breakout in Philadelphia killed some 5,000 people. Ffirth studied the disease a few years later at the University of Pennsylvania, and became convinced that it was not actually contagious. Then only sensible way to prove this was, of course, to make cuts in the flesh of his arms, and then rub vomit from a yellow-fever patient into those cuts. When that didn't give him the disease, he stepped up his game by pouring yellow fever puke into his eyeballs. When that didn't do the trick, he sautéed some yellow fever puke in a pan and inhaled the fumes. Then he just straight up drank it. Oh, and then he moved on to blood and urine. He declared his research definitive proof that yellow fever was not contagious.

He was correct . . . provided the patient in question had late stage yellow fever, which we now know is no longer contagious.

The takeaway here is something like: If you were to, say, drink and rub and basically freaking vape a bunch of puke from exclusively late-stage yellow-fever patients, you would likely come out fine. Except for, you know, drinking all the puke and creating misguided science.

BITTEN BY THE EPIDEMIOLOGY BUG

Cuba, 1869

Eighteen-year-old Walter Reed was (and still is) the youngest person in history to earn a medical

degree from the University of Virginia. When his youth proved to be an obstacle in job-hunting, he decided to join the Army. His first assignment the unenviable task of trying to figure out why so many U.S. soldiers in Cuba were dying from yellow fever; the suspected vectors included waterborne illness, or person-to-person contact. Reed didn't really agree with either theory. His money was on the idea, proposed by Cuban doctor Carlos Finlay, that the infection was spread by mosquitoes.

To test this theory, Reed assembled a team of infectious disease specialists to investigate this hypothesis. These doctors—James Carroll, Aristides Agramonte, and Jesse W. Lazear—volunteered to be bitten by mosquitoes that had just bitten yellow-fever patients in a controlled environment.

Carroll got sick but got better. Lazear got sick and . . . didn't. We're not actually sure what happened to Agramonte.

These brave research volunteers helped Reed's team to conclusively prove that yellow fever is indeed spread by mosquitoes. Reed got a hospital named after him, even though—as we're betting that some of his colleagues pointed out—he never got bitten himself.

THEY DON'T MAKE LAB ASSISTANTS LIKE THEY USED TO

Kiel, Germany 1898

Very few could deny that the German surgeon August Bier's experiments with spinal anesthesia constitute a massive contribution to the benefits of modern medicine.

And certainly nobody can say that the guy didn't know how to party.

Bier first tried his revolutionary procedure, which he called "cocainization of the spinal cord," on a laborer who had previously had a bad reaction to general anesthesia of the time. The treatment was a success in that it performed what we'd now call the spinal block, keeping pain signals from reaching his brain, which allowed him to remain conscious throughout minor surgery. There were a few kinks to work out, of course—specifically, a severe headache that followed the spinal puncture.

> Sorry, I just have to take a quick moment to recognize how radical the word "cocainization" is. I'm not a proponent of drug use, but it does make me a little sad that none of my work will ever be described as "cocainized."

Bier decided to try being on the receiving end of a spinal anesthetic administered by his assistant, August Hildebrandt. Somehow, they attempted the procedure using a needle that didn't properly fit the syringe, resulting in Bier losing a significant amount of cerebrospinal fluid; the anesthesia didn't take effect. Hildebrandt immediately offered up his own spine to his mentor and received a much more effective dose of anesthesia. (How effective? Turn the page for a timeline of how these party animals spent the rest of their night, straight from Bier's own writing.)

STRAIGHT TO THE HEART OF IT

Ederswalde, Germany 1929

In the early days of the 20th century, most folks were pretty sure that sticking stuff in your heart that didn't belong there (that is to say, anything that was not already there) was a pretty good way to instantly die. German doctor Werner Forssmann disagreed. He had this wild idea that you could pass a catheter through the body's venous system to the heart to deliver meds, measure blood pressure, and inject dye to look at the heart in an X-ray.

In 1929, he came up with a way to show those naysayers—by jamming a foreign implement into his own heart. His hospital's department head was not thrilled with this idea, but that didn't stop Werner. He went straight to the operating-room nurse in charge of a crucial locked supply

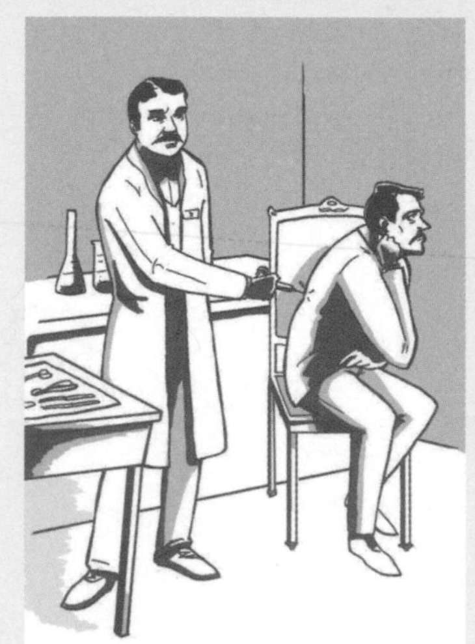

The Great Spinal Cord Experiment

Experiment Begins

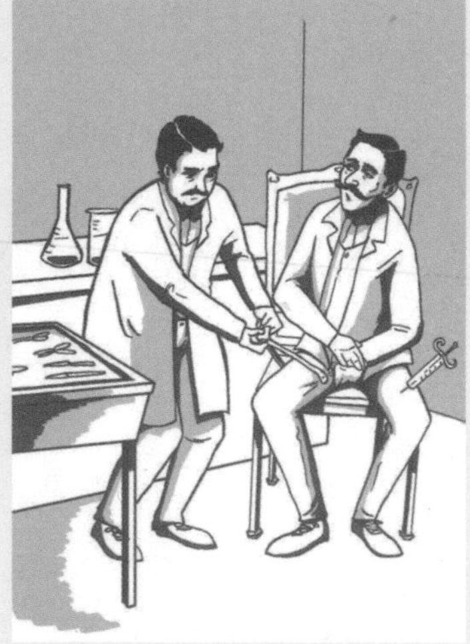

After 10 minutes
A long needle was pushed down to the femur without evoking the least pain. Pinching the skin severely and seizing and crushing it in toothed forceps was experienced as pressure.

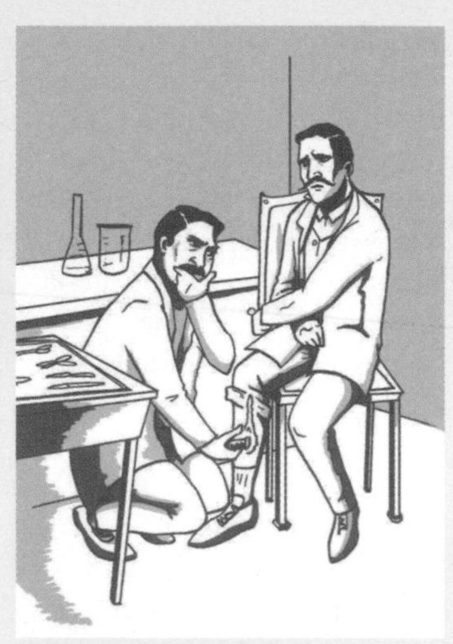

After 13 minutes
A burning cigar applied to the legs was felt as heat, but not as pain. Ether produced a feeling of cold.

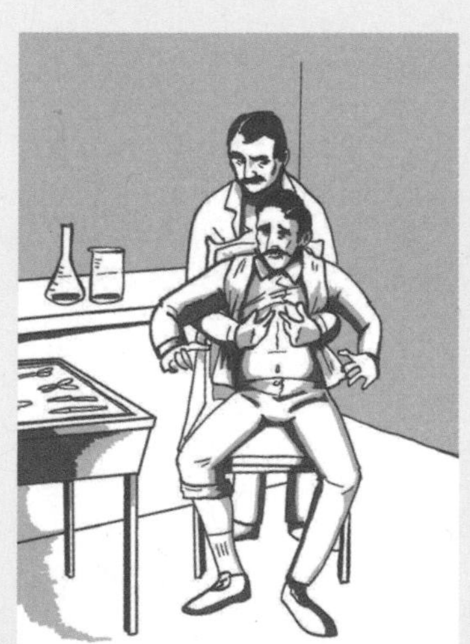

After 18 minutes
Strong pinching was hardly felt at all below the level of the nipples.

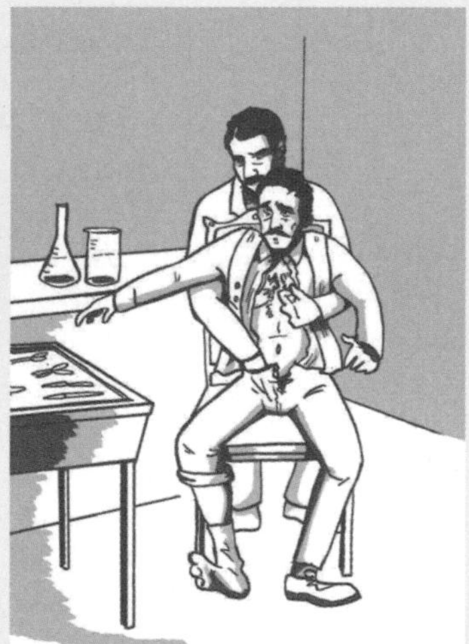

After 20 minutes
Avulsion of pubic hairs was felt simply as elevation of a fold of skin, but avulsion of chest hair above the nipples was very painful. Strong hyperextension of the toes was not unpleasant.

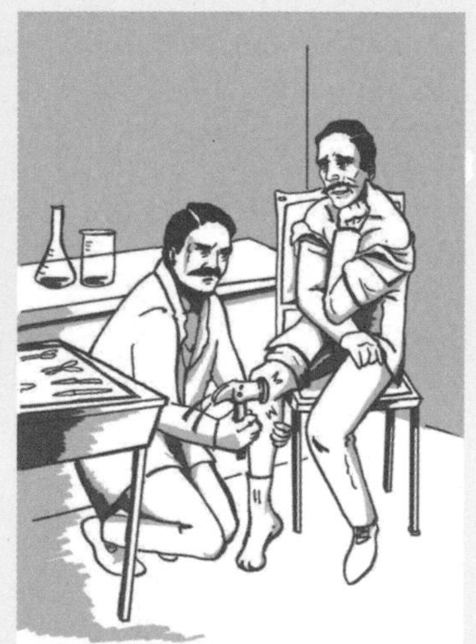

After 23 minutes
A strong blow to the shin with an iron hammer did not provoke pain.

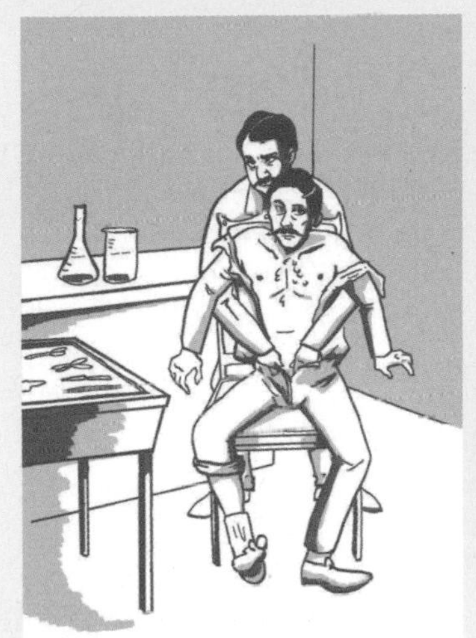

After 25 minutes
Strong pressure and traction on the testicles was not painful.

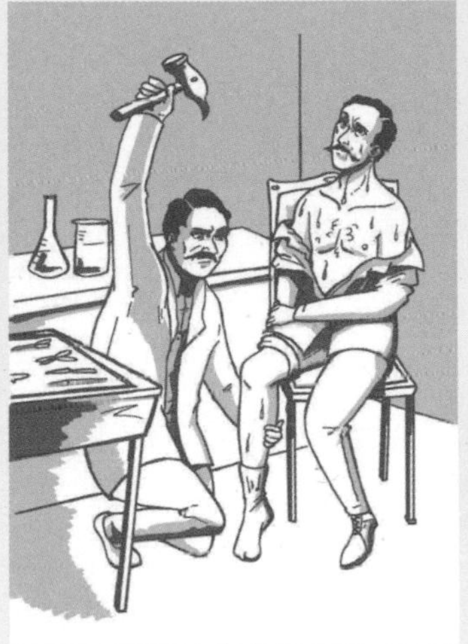

After 40 minutes
Strong blows on the shin did not hurt. The entire body began to perspire gently.

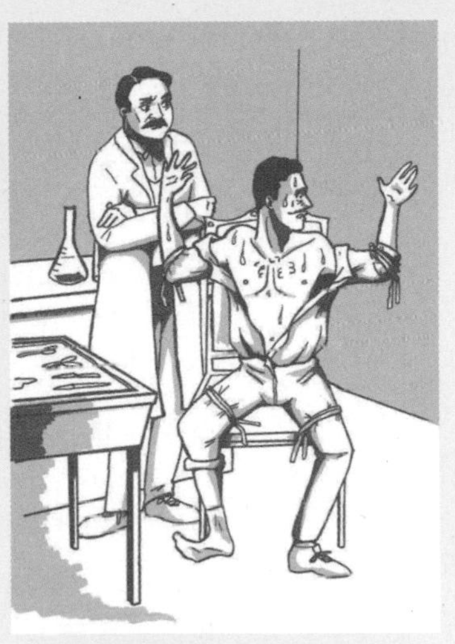

After 42 minutes
Constriction by a rubber tube tourniquet around the thigh produced no pain, but around the upper arm was very painful.

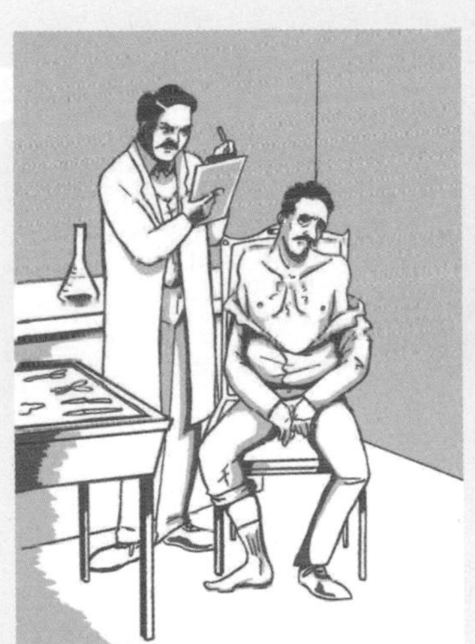

After 45 minutes
Pain sensibility began to recover, but was still considerably obtunded. Sensibility was gradually recovered completely.

Once the drugs wore off, Bier reported that they celebrated over a nice dinner, accompanied by wine and cigars.

Hildebrand experienced headaches for four days. Bier, oddly enough, was on bed rest for nine days. Maybe he pulled a muscle swinging that hammer.

cabinet and asked her to assist. She agreed, but only if he gave up the notion of operating on himself. She would make that possible by volunteering as a human guinea pig. Forssman pretended to acquiesce to this idea, going so far as to strap her to the operating table and numb her wrist. And then he went ahead and passed the catheter through his own antecubital vein and a good distance toward his heart.

He then unstrapped the nurse and the two of them calmly walked downstairs to radiology, where he used fluoroscopy to advance the catheter into his right ventricle.

Or, in non-medical journal language, he stuck a long, pointy tube into that crook-of-the-arm vein doctors like so much, and pushed it through his veins like a subway train through a tunnel, trying to get it all the way to his heart. To finish the job, he and the apparently pretty chill nurse went and used the hospital's X-ray machine to check on where the tube was in his circulatory system, and shove it through the rest of the way the Heart Central Station.

Also, he didn't drop dead as a result of this little escapade.

This strange story is about to get stranger. Forssman's peers did not generally approve of this stunt, its success notwithstanding. So, he sought out a less ethical group of physicians. In Germany. In the 1930s. (Spoiler alert: it was the Nazis.) He joined the Nazi party, eventually rising to the rank of major before he was captured by the Americans in 1945. Oh, and eleven years after that, he won the Nobel Prize in medicine for his work on cardiac catheterization.

We know—there were a lot of ups and downs with that one. Catch your breath.

THE WORMS CRAWL IN

Nottingham England, 2004

You may hate allergies, but how much? Oh, sure, you may have Claritin empties piling up, and an Afrin habit so vicious you have to hide it from your family. But do you hate allergies as much as biologist David Pritchard does? We suspect that you do not.

While working in Papua New Guinea in the 1980s, Pritchard noticed that locals who were infected with hookworms were rarely bothered by immune system disorders, such as asthma. He figured there was a connection, but it took him years to formulate a solid theory. Our bodies tend to react badly to foreign intruders, but hookworms show up, move in, and somehow keep from getting expelled—which implies they can shut down your natural immune responses.

Since allergies are, essentially, an overactive form of immune response, Prichard wondered if that hookworm superpower could be harnessed and used for good.

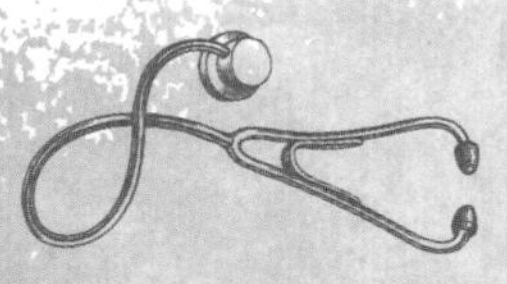

Sydnee's Fun Medical Facts

Dr. Forsmann's DIY approach to that Nobel Prize–winning discovery isn't as unprecedented as you might guess. A study in the Texas Heart Institute Journal examined 465 cases of self-experimentation over a period of fifty years and found that twelve of those brave souls went on to win a Nobel Prize for those experiments, which is a pretty good number all considered.

He actually worked on this theory for almost twenty years, a rarity in the field of self-experimentation, which seems to favor the "hold my beer" approach to experiment design. That said, here's how much of a hate-on Pritchard had for allergies: In order to prove this theory merited further investigation, he hatched a whole lot of hookworm larvae on a piece of cloth, tied it around his arm, and left it there for days until he was certain they had fully . . . infiltrated.

"The itch when they cross through your skin is indescribable," he told the *New York Times*. "My wife was a bit nervous about the whole thing."

The results were encouraging and have inspired further study into what's now known as "helminthic therapy." Though there's plenty of work to be done before hookworms become a standard part of your doctor's toolkit, some folks aren't waiting to start disrupting your innards. At least, that was the case of a Silicon Valley entrepreneur who graduated from running an online helminthic therapy discussion group to founding a clinic in Mexico where, for only $3,900, you can experience his self-designed (and unregulated) hookworm-based "treatments."

Sadly, that particular clinic isn't an anomaly: Helminthic therapy is on the rise. Despite the stomach cramps, diarrhea, and fatigue that ensues, many desperate patients have started giving it a go for a variety of autoimmune and allergic conditions.

While, yes, there are some encouraging results, there's just not strong enough evidence of its effectiveness to justify the risk. Also—and we should have mentioned this first—it's completely illegal, so you can't be completely certain that the black-market worms you're buying are even what the seller says they are.

If this isn't enough to discourage you from giving this a shot, consider the name of the hookworm you may be ingesting: *Necator americanus*. For those of you who don't speak Latin, that means "American killer."

DOES THIS STILL HAPPEN TODAY?

Self-experimentation will exist as long as there are medical questions to be answered, and a lack of sufficient grad students or other, to help scientists find answers. But the whole DIY thing is definitely on a downturn, historically speaking.

Of the 464 cases of self-experimentation documented by the Texas Heart Institute Journal study Sydnee mentioned earlier, 189 cases occurred in the first half of the 20th century, but only eighty-two took place in the second half.

Maybe all of the dabblers are getting discouraged, leaving the field to mavericks like Pritchard. Or like Australian internist Barry Marshall, who proved modern theories about what causes ulcers to be incorrect. All it took was a nice lunch of *Helicobacter pylori* broth to prove that ulcers are caused by a transmissible organism, and not by, say . . . jalapeño nachos. He won the Nobel Prize too, by the way.

See, that's the interesting thing about self-experimentation: It works. Usually. According to that same THIJ study, "in a remarkable 89 percent of instances, the self-experiments obtained positive results in support of a hypothesis or valuable data that had been sought."

Is it a case of the medical gods favoring those bold enough to perform science on their own personal bodies? Or maybe there's an even more obvious explanation: Before you start injecting yourself with weird crap, you should really try to be extra darn sure that it's going to work.

So What's the Deal With: HOMEOPATHY?

Looking to cure that illness? Try this toxic herbal mixture diluted until it's basically just pricey water. That'll work . . . won't it?

When Did This Become a Thing?

Homeopathy traces its roots all the way back to ancient Greek medicine. In 400 BCE Hippocrates prescribed a small amount of mandrake root as a cure for mania, based in the theory that "like cures like"—if a lot of mandrake can cause mania, the thinking went, then maybe a little bit can cure it? This became known as the Doctrine of Signatures; and served as the foundation for an entirely new medical tradition. But how did an ancient practice see a modern revival? Enter Christian Friedrich Samuel Hahnemann.

Born in 1755 in Meissen, located near Dresden, Hahnemann was—in the parlance of the time—super-duper-smart. His father, disdainful of formal education, took him out of school for"thinking lessons," where he would instruct the boy to simply sit there and . . . uh . . . think. Despite this unusual practice (or maybe because of it; who knows?), Hahnemann quickly mastered pharmacy, botany, physics, and at least ten languages. After working as a translator, he studied medicine at Leipzig and Vienna, and at last graduated from the University of Erlangen.

Late-1700s medicine was a strange and wild place. There were no shortage of theories and research, and no shortage of bizarre experiments and treatments. This was the Age of Heroic Medicine, when treatments were prescribed in order to balance out the body's humors, with plenty of sweating, bloodletting, and vomiting—and what with that being the easiest way of balancing the humours, this probably meant a lot of vomiting. (It did mean a lot of vomiting.)

As a relatively new arrival to medicine, Hahnemann took a look at all this and said, "Y'know, I *think* we're doing more harm than good here." He advocated clean living, a proper diet, fresh air, and exercise as the basis of good health, which, actually, we are pretty on board with. Ultimately, though, he became frustrated by the unscientific treatments of other doctors, and took up translation again. While translating William Collins' *Materia Medica*, he came across a passage describing cinchona tree bark as a cure for malaria and, out of curiosity, tested it on himself. Although the bark contained the anti-malarial quinine which was, indeed, a treatment for malaria, Hahnemann actually developed symptoms of the disease after taking it.

Unaware that this was the result of cinchonism (a quinine overdose) he erroneously followed the same logic as the ancient Greeks, and decided to test small amounts of poisonous herbs and animal venom on himself, his friends, and even some of his eleven children. (Good thing he had some to spare, huh? We kid!)

Yet, when those substances were given in small amounts to already-ill folks, those people got even sicker. Surprise! So, he quit! No, no, he went further still: Hahnemann figured that even smaller amounts were needed—1 drop of an herbal extract in 99 drops of water or alcohol, then a drop of that diluted to one-hundredth, and again, and again. Realistically, no measurable or effective amount of the substance remained, except for a "footprint" or "echo," and this diluted mixture—essentially water or alcohol—was given to a patient. Hahnemann named his idea using the Greek words *hómoios*, "like," and *pathos*, "suffering," calling the practice "Homeopathy."

Hahnemann, who considered plenty of other doctors to be unsafe and eccentric, was himself seen as a quack and controversial by many. Regardless, he received the patronage and protection of the archduke of Saxony, so what did he care? He remarried eventually, and moved to Paris, before he died at the age of ninety.

Gemahlt von Schoppe 1831. Stahlstich v. Leop. Beyer in Wien.

So, What Happened Then?

In the 1820s, a Dr. Constantine Hering was studying Hahnemann's theories with a goal of scientifically refuting them. While dissecting a cadaver, Hering badly cut his finger, and the ensuing gangrene lead many other physicians to advise amputation. Less than thrilled with this plan and feeling desperate, Hering decided to give homeopathy a shot, and applied a concoction which contained some small dilution of white arsenic. Much to his surprise, the wound healed and he got to keep the digit. Spurred on by this recovery, Dr. Hering became a convert, and helped found the Homeopathic Medical College of Pennsylvania in 1848 (now part of Drexel University).

By 1900, there were 111 homeopathic hospitals and twenty-two medical schools focused on the practice of homeopathy, but the discipline's fall was as sudden as its rise. With improved understanding of the science of medicine, homeopathy quickly fell out of fashion, and by 1923, only two schools in the United States were still teaching it.

There is zero evidence to support the use of homeopathy. Homeopathy can be dangerous, in that it keeps people from seeking out actual medicine, and it's also a rip-off. The fact that homeopathic cures are sold alongside actual medicine is one of the dumbest things that happens on Earth, and we're writing this in 2018, so let's just say that it's a heated competition.

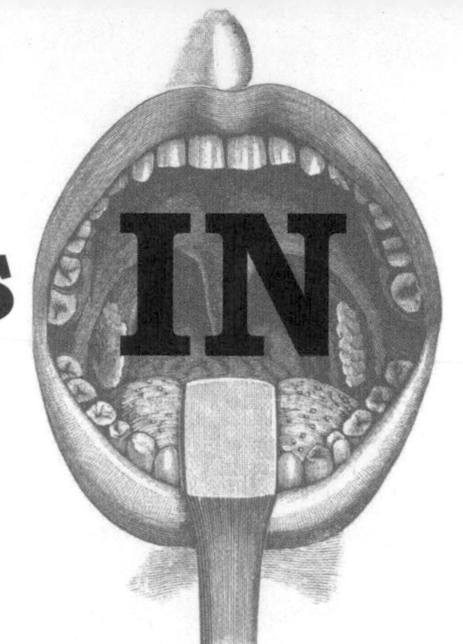

THE DOCTOR IS IN

It's time to take another break from rubbernecking at the flaming wreckage on the medical-history highway to answer real questions from real listeners of *Sawbones*, a real podcast.

If you knew that there was only one antibiotic which would work on a patient, but they were allergic to it, would you still prescribe it?

Sydnee: This is a very unlikely scenario. There are many different classes of antibiotics and most bacteria are not resistant to all but one antibiotic. That being said, if you're about to knowingly expose someone to they're allergic to, medication or otherwise, you can pretreat them with certain medications that will help decrease the severity of the allergic response.

There's a similar situation we may be more likely to find ourselves in, with patients who are allergic to contrast dye. Even if you have that condition (it's rare; less than one percent of patients have severe reactions), you may still find yourself in need of a CT scan, so we'd have to pretreat you to limit the danger.

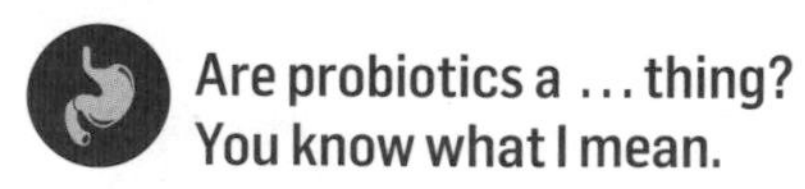

Are probiotics a . . . thing? You know what I mean.

Sydnee: Yes. Put simply, probiotics are "good bacteria." They're part of the normal flora that exist in your gut and aid in digestion. Not everyone needs to take probiotics all the time, however. If there's a concern your gut flora has been disrupted by illness or antibiotics, probiotics can help restore the balance.

If you have symptoms such as diarrhea or constipation, after a round of antibiotics or disease affecting the digestive tract, it may be a sign that you could benefit from ingesting probiotics. Any yogurt with live active cultures can help repopulate your gut with a bunch of body-friendly bacteria.

Justin: So you're saying I don't necessarily need to buy the kind that helps Jamie Lee Curtis poop, and I hear you. But if it's all the same, I'm going to stick with the kind that helps Jamie Lee Curtis poop.

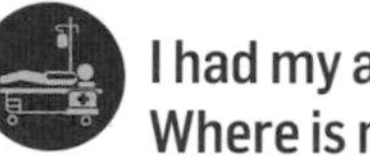

I had my appendix removed. Where is my appendix now?

Sydnee: I don't know! It's probably in the trash somewhere, I guess. There are special disposal methods for organs, biohazard containers, that sort of thing, but it was still headed for the trash. The pathology lab would have examined it, and they probably took a sample to check for disease. After that, though? Right in the trash. Unless it was really weird and we needed to do research on it. Wait! I just remembered, I know this: We incinerate them. So it's nothing but dust by now. Probably.

Justin: Wow . . . I'm just . . . spellbound. Syd, you really do know how to make medical lore come to life.

Are there any essential oils that actually do something?

Sydnee: There's some reason to suggest that aromatherapy could help with stress, anxiety, and depression, especially when administered via massage. Peppermint oil can help with nausea and vomiting during labor. After that, the evidence gets a lot dicier. There are some small individual studies that indicate they may help with pain and insomnia.

Fennel, aniseed, and sage may help with PMS and menopause because they have estrogen-like compounds, but we need more research to prove that they have any measurable effect in humans. And it's a known phenomenon that a person's belief in the benefit of any given therapy influences whether or not they'll respond to it. So, in short, the evidence is threadbare, and anyone who suggests differently is probably not someone you should take medical advice from. Feel free to inhale the aroma of oils you like, and use them in a massage, but don't eat them. Also, stay away if you have allergies or asthma, or if you're pregnant.

Justin: It works better if you believe in it? I'm not a medical professional—that's been pretty well established at this point—but anything that has the same amount of efficacy as Dumbo's magic feather doesn't fill me with a ton of confidence.

Is it true that yogurt douches can prevent yeast infections?

Sydnee: Yogurt can help prevent yeast infections and diarrhea that you get from taking antibiotics, by providing good bacteria to help fight off the invaders. We discussed this a few questions ago; what we didn't discuss is whether or not putting yogurt into your vagina is a good way to use probiotics. So let's discuss: Don't do that.

Also, never douche. It kills even healthy bacteria, and can alter your pH balance, thus leading to infection. Douches are a creation of the patriarchy. Let vagina be vagina.

I've heard the pins and needles feeling when parts of your body fall asleep is caused by lack of blood flow, but why exactly does it feel like that?

Sydnee: The sensation of our limbs "falling asleep" is actually related to an interruption in the signals between our nerves and our brain. Pressure on our nerves directly can interrupt these signals, as well as pressure on the surrounding blood vessels that deliver oxygen to our nerves. So, in a sense, blood flow is part of the problem, but its the nerves that are actually causing the "pins and needles" sensation. These feelings are also called paresthesias. As the pressure is relieved and feeling starts to return, it can take awhile for the signals to communicate correctly. In the meantime, the sensations are the result of impulses starting to emerge from the nerve roots. In conclusion, just move around and shake the limb, and you'll be fine.

Justin: Or, you can try my patented coping method: complaining how uncomfortable it is—until your wife slowly points at her C-section scar while she stares at you silently. Mine always seems to clear right up after that.

THE AWESOME

Wait, wait, it's not all doom and gloom, we promise! Look, we even wrote an inspiring sonnet!

And though we may seem lost here in the dark
It's not all blood and guts (though that's still fun)
The heroes get a chance to leave their mark
Like Salk who wouldn't patent his new sun

So poison we will drink if fate demands
We'll bleed ourselves to find the truth within
The vinegar perfume still on our hands
We eat the fudge we pray is medicine

Sometimes we human beings got it right
As hard as that may be now to believe
Maybe by luck sometimes we find the light
And accidental greatness we achieve

So let us end on this last happy note
Except Kellogg who was kind of a scrot(um)

The Poison Squad

And you thought your job was hard. Maybe it is, but how much rotting food does your boss make you eat?

If you have a food-poisoning-related problem, if no one else can help, and if you can find them . . . and if they're not too full or barfing too much . . . maybe you can hire the Poison Squad.

Yes, we're going to hear the story of some guys who probably had one of the roughest, yet coolest-sounding, gigs in medical history, we promise. But first, we need to take a moment to talk about food poisoning.

We've known that food can be a potential threat since ancient times. Food poisoning has been known throughout history as "Death in the Pot," and scientists have found evidence of various foodborne illnesses in mummies, bog bodies, skeletons, and of course, coprolites. Justin—before you ask—that's fossilized feces.

Okay, I just feverishly searched the Internet, and as far as I can tell, there's never been a punk band called "The Coprolites." So I'm calling dibs right now. DIBS. There, it's in a book, so that's the law. So . . . uhh . . . do any of y'all play drums?

TUMMY TROUBLES OF THE ANCIENTS

In the beginning, we probably figured out that food caused illness, as well as which foods to avoid through trial and error, and by watching what foods animals avoided. And then of course, according to the Old Testament, God told Moses which animals were "clean" to eat, as well as what could be considered food-safety tips—which we now know as the Jewish dietary rules.

Hippocrates noted that clean water tasted better, and so he developed a water filter that he used for himself and for his patients as well. He even boiled water to clean it, which was pretty revolutionary for the time considering that he would have had no knowledge of any microorganisms or substances that he couldn't see or smell in the water.

The ancient Greeks also knew better than to eat any animal that was diseased—we've known to avoid spoiled meat since prehistory. This did not, as some have erroneously believed, lead to people spicing spoiled meat to disguise its flavor and pass it off as fresh. In fact, there are laws dating back hundreds of years written against adulterating food and drink to make it appear safe, or more appealing, to sell it. Pliny the Elder even warned of wines that were tainted with noxious herbs to look brighter and more flavorful.

While we used our noses to ferret out possibly dangerous food and drink for many centuries, we didn't know why those things made us sick. That mystery didn't begin to unravel until the microscope came along in the 1600s, allowing anyone interested to see all the secret little germs hiding beneath our noses. After that discovery, it was another couple hundred years until Louis Pasteur recognized that those little germs in food probably caused illness, and developed heating food, or pasteurization, to kill them.

DON'T LICK THAT IGUANA

There are many examples of foodborne illness, but we have to start somewhere. So, let's talk about the one we isolated and blamed for

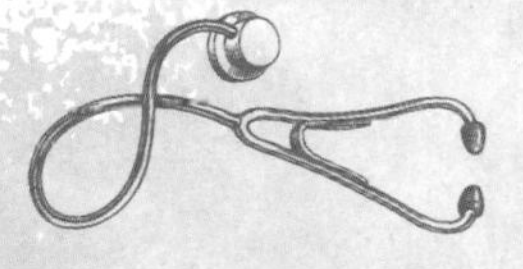

Sydnee's Fun Medical Fact

Hiding spoiled meat with spices makes even less sense when you consider just how wildly expensive those spices were in ancient times. According to economics professor John Munro, a pound of ginger in ancient Rome would have cost roughly 5,000 days' wages. Anybody who could afford that could, obviously, afford to just buy some new meat.

puking first: *Salmonella*. This is a bacteria in the *Enterobacter* family, and is most famously responsible for typhoid fever. Well, okay, "famously" is a word some coauthors of this book might use, and then immediately realize why people think they are unpleasant dinner company. There are multiple strains of *Salmonella* bacteria, but *enterica* is the one derived from underdone chicken—also known as "the one you're probably terrified of if you've ever watched a Lysol commercial." While it's true that poultry is a big culprit, so are reptiles. Not eating them, necessarily, so much as touching them a lot. (Specific reptiles to keep at arm's length include the green iguana and the red-eared slider turtle.)

This rather unfortunate side effect of reptile ownership actually led to the U.S. FDA issuing what was called the "four-inch law" in 1975. It dictated that any turtles sold in the United States had to have a carapace length of at least four inches, because then it would be harder for a kid to stick the turtle in his or her mouth.

That's only for parents who are raising quitters. I'm gonna teach our girls they can fit any size turtle in their mouths if they just believe in themselves.

The symptoms of food poisoning are fairly well known to us all, but in case you are the single lucky human who has never experienced it, you can expect nausea, vomiting, abdominal cramping, and diarrhea. It starts about twelve to seventy-two hours after you become infected, and lasts about four to seven days. Usually, there's not much to do in terms of treatment, other than to stay hydrated and ride it out.

Nobody wants to experience that, but nobody wants to throw food out either, right? So, humanity started fighting back against food spoilage.

NOW WITH EXTRA STRYCHNINE!

Heating and cooling were simple enough ways to safeguard food in the home, but what food manufacturers really needed were ways to preserve food for sale. Great idea . . . except no one had any idea what was helpful or safe in terms of chemical additives. So, in the early 1900s, bacteria got a new partner in their never-ending quest to make food dangerous: humans. Many packaged goods ended up riddled with preservatives that had never been tested for safety. Pesticides were being used on produce without any thought of what they may do in human bodies. Chemicals such as borax, formaldehyde, and strychnine were added to meat to preserve it, and some foods were stuffed with fillers, including chalk or sawdust, to fool the consumer into thinking it was fresher or more wholesome.

These unregulated products obviously didn't have modern-day-style labels to warn the consumers of what may be inside, and these rogue preservatives were wreaking havoc on the health and safety of the American people. The problem was that no one knew how to separate the hazardous preservatives and pesticides from the harmful ones. But who was willing to try them out and see what was making everyone sick?

DR. WILEY TO THE RESCUE

Doctor Harvey Wiley was up to the task. Well, not so much the man himself—he wasn't about to eat potential poisons—but he was willing to live with the guilt of convincing other people to do it. We ask you, dear reader, isn't that true heroism?

A medical doctor from Indiana with additional training and education in chemistry and food science, Dr. Wiley was appointed to the position of Chief Chemist at the U.S. Department of Agriculture in 1882. Initially, he was working mainly on sugar chemistry and investigating sorghum as a possible sugar source for the United States, but by the turn of the century, food safety was taking center stage. Congress wanted to know if the various compounds we were putting into foods to preserve them were safe. So, in 1905, they gave Dr. Wiley $5,000 to figure it out. He took the money and created what he termed the "poison squad."

A note on Dr. Harvey Wiley: He was smart, worked hard to improve food safety, and was a good scientist. It bears mentioning, though, that he was also a terrible chauvinist. Dr. Wiley believed that women lacked the brain power of men, and refused to allow them to participate in any aspect of the process. Women were not even allowed to serve as cooks, as he felt that they were too dumb to even poison correctly. He was also fired from a position at Purdue University for riding a bicycle, but that seems less egregious.

I briefly floated the idea to Sydnee that the previous paragraph could have been a "fun fact." It . . . did not go well.

The first thing Dr. Wiley needed to do was find some patsies . . . uhh, willing subjects. He advertised for "twelve young clerks, vigorous and voracious." Once people began to respond, he had an in-depth screening process for them to pass. They all also had to pass the civil-service exam and demonstrate that they were of "high moral character."

In addition, interviews with family and friends ensured that all of the volunteers had reputations for "sobriety and reliability." The lucky chosen would be asked to pledge for a year to only eat what was given to them. Even if it had, you know, poison in it.

The surprising thing? It worked. Men were actually eager to volunteer for something called "the poison squad." Not only did they volunteer, but they sent letters to Dr. Wiley practically begging him to pick them for his experiments. Here is an example of one such letter:

Dear Sir:
I have read in the paper of your experiments on diet. I have a stomach that can stand anything. I have a stomach that will surprise you. I am afflicted with 7 diseases. Never went to a doctor for 15 years. They told me 15 years ago that I could not live 8 months. What do you think of it? My stomach can hold anything.

So, the doctor gathered his subjects, and the tests commenced. Before dinner, as is customary at all the finest restaurants, the diners' vital signs were taken, and they submitted some, well, samples.

She means pee-pee and poo-poo. And maybe blood.

It was all a very elegant affair, twelve well-dressed young men sitting at nicely-set tables dining together. The men were all healthy

(at the outset), and over the course of what would eventually be five years, they were fed various meals laced with borax, sulfuric acid, saltpeter, copper sulfate, and formaldehyde.

DINING ON DISASTER

Over time, the squad members were given higher doses of each poison and then carefully examined. The doctor recorded their weights; took blood, urine, and hair samples; and monitored their vital signs carefully. The men reported their symptoms as well, especially as they increased in severity with escalating doses. Inevitably, they got pretty sick, and some tried to drop out. The doctor had to bargain with the squad to get them to finish out the borax trial, which was especially brutal.

Over the course of five years, the tests continued with different volunteers as needed. Each of the possible poisons were tried, and the squad took the challenge as long as they were able. By 1907, many members were described as being "on a slow approach toward death." The experiments ended, the findings were presented to Congress (and the media) and the outrage that ensued led to the Pure Food and Drug Act of 1906. This ensured that consumers would be informed as to what was in the food they bought, and that preservatives and additives had to be held to certain safety standards.

It was so closely linked to the trials of the poison squad that the Pure Food and Drug Act was initially known as the "Wiley Act." Eventually, though, Roosevelt took so much credit for it that Wiley lost his top billing.

At least he could comfort himself with the knowledge his work lead to the creation of the Food and Drug Administration, food safety guidelines and his eventually being known as "the father of the FDA."

POISONOUS MELODIES

How much did people love the poison squad? It was the subject of at least two different songs, the lyrics of which you'll find reprinted below:

Respectfully Dedicated to the Department of Agriculture By S. W. Gillilan

O we're the merriest herd of hulks
that ever the world has seen;
We don't shy off from your rough
on rats or even from Paris green:
We're on the hunt for a toxic dope
That's certain to kill, sans fail.
But 'tis a tricky, elusive thing and
knows we are on its trail;
For all the things that could kill
we've downed in many a gruesome wad,
And still we're gaining a pound a day,
for we are the Pizen Squad.
On Prussic acid we break our fast;
we lunch on a morphine stew;
We dine with a matchhead consomme,
drink carbolic acid brew;
Corrosive sublimate tones us up
like laudanum. ketchup rare,
While tyro-toxicon condiments
are wholesome as mountain air.
Thus all the "deadlies" we double-dare
to put us beneath the sod;
We're death-immunes and we're proud
as proud—Hooray for the Pizen Squad!

As sung by Lew Dockstader—in His Minstrel Company, Washington, D. C., week of October 4, 1903.

If ever you should visit the Smithsonian Institute,
Look out that Professor Wiley doesn't
make you a recruit.
He's got a lot of fellows there that tell
him how they feel,
They take a batch of poison every time
they eat a meal.
For breakfast they get cyanide of liver,
coffin shaped,
For dinner, undertaker's pie, all trimmed
with crepe;
For supper, arsenic fritters, fried in
appetizing shade,
And late at night they get a prussic
acid lemonade.

(Chorus)
They may get over it, but they'll never
look the same.
That kind of a bill of fare would drive
most men insane.
Next week he'll give them moth balls,
à la Newburgh, or else plain.
They may get over it, but they'll never
look the same.

Unsatisfied with being a scientific pioneer, chauvinist, and cycling rebel, Wiley was also a poet. Here's his ode to food safety.

We sit at a table delightfully spread
And teeming with good things to eat
And daintily finger the cream-tinted bread
Just needing to make it complete.

A film of the butter so yellow and sweet
Well suited to make every minute

A dread of delight.
And yet while we eat
We cannot help asking "What's in it?"

Oh, maybe this bread contains alum and chalk
Or sawdust chopped up very fine
Or gypsum in powder about which they talk
Terra alba just out of the mine.

And our faith in the butter is apt to be weak
For we haven't a good place to pin it

Annato's so yellow and beef fat so sleek
Oh, I wish I could know what is in it.

Listen, friend, we're starting to feel really close to you after the past hundred-and-some pages, so we're just going to come right out and say it: We think you're getting a little bit full. Listen to yourself: you're just kinda . . . sloshing around. Haven't you put it off enough? Isn't it time to get some of that blood out of there?

Oh sure, you could donate it—*like a sucker*—but the important thing is that we get that excess blood out of there and get you on the road to recovery.

For almost as long as we've been trying to fix people, we've been doing—there's no nice way to say it—*just* about the most unintuitive thing conceivable: bloodletting, the practice of draining some of a patient's perfectly fine blood in the hopes that it'll fix what's ailing them.

Maybe even more shocking than the practice itself (well, equally shocking at least; bloodletting is *wicked* stupid) is how many things we've prescribed bloodletting for.

Seizures

In the late 1600s, English king Charles II suffered a seizure, but luckily his crack team of physicians knew exactly what to do: drain him like a Capri-Sun. They took sixteen ounces from Charles II, followed up with a bracing round of enemas. Chuck 2 had more seizures, which is *weird* because of all the blood they took, but luckily they had one last Hail Mary in the playbook: taking a bunch more blood. By the time he had died (from extremely related causes), they had drained him of twenty four ounces; that's two Diet Coke cans of blood. (Sorry for making any vampires out there super thirsty.)

Fevers

Did you know leeches can drink ten times their own weight in blood? Well, early 19th century French physician Francois Broussais sure did! He believed any fever in your body could be fixed by tossing a leech on it and letting it remove the excess organ inflammation. That . . . isn't right. You knew that wasn't right, right? Right.

That didn't stop the Parisians, of course. By the 1830s, five to six million leeches a year were used in the city alone, with another thirty million leeching blood across the country per annum. Passion for the practice had waned by the end of the century, so if you wonder why you always see so many leeches panhandling in Paris, now you know.

Yellow Fever

Benjamin Rush was an unorthodox, if formative, physician in the late 1700s and early 1800s. (He signed the Declaration of Independence. He gave Lewis and Clark mercury pills to help them poop. He was a big deal.) He also shared some of Broussais' unconventional ideas about fever. In 1815, he went to bat for them, specifically with regards to yellow fever. He wrote:

"I have attempted to prove that the higher grades of fever depend upon morbid and excessive action in the blood-vessels. It is connected of course, with preternatural sensibility in their muscular fibres. The blood is the most powerful irritant which acts upon them. By abstracting a part of it, we lessen the principal cause of the fever. The effect of blood-letting is as immediate and natural in removing fever, as the abstraction of a particle of sand is, to cure inflammation of the eye, when it arises from that cause. [Bloodletting] imparts strength to the body, by removing the depression which is induced by the remote cause of the fever."

Yup, that's right, *the blood was the problem the whole time.*

Assassination

Okay, admittedly this one is a bit of a stretch. When George Washington took ill after a snowy ride in 1799, his doctors recommended bloodletting. That wasn't that weird at the end of the 18th century, but what was *slightly* more unorthodox was how much they drained . . . eighty ounces. That's just under *half* of his blood. You can't live without half your blood, and America's first president was no exception.

We're not saying it was intentional, but it's hard to argue that doctors didn't technically assassinate Washington. Is this already the plot of an *Assassin's Creed* game? If not, we call dibs on it.

Death by Chocolate

That's just the name of one of those over-the-top brownie desserts, not a real thing. Because chocolate is good medicine.

Well, we're not sure about that, and you haven't even read this chapter yet. But let's just all try to stay positive, okay? How could such a wonderful thing as chocolate not have marvelous healing powers? It just makes sense.

Are you familiar with *The Secret*? You know, the Laws of Attraction? Like attracts like? Put simply, the idea is that the universe will deliver anything to you provided you just visualize it hard enough. So let's all start putting it out into the universe now that chocolate is medicine, okay? Because we really, really need chocolate to be medicine. You hear that, Universe? Chocolate is medicine.

Please keep repeating this mantra as you read, we're not sure how close attention the universe is paying (judging by recent evidence, the answer is not particularly).

FOOD OF THE GODS

Let's begin with the history of chocolate itself. No, it's not technically medicine, but we've been writing about a lot of blood and poop and stuff—just give us this brief reprieve, okay?

Humanity has been brewing up fermented chocolate beverages since around at least 1400 BCE when people in Mesoamerica first domesticated the cacao tree and began making drinks from its fermented beans. Mayans thought the drink had a connection to the gods. By 1400 CE, Aztecs were using the beans as currency.

In these early American cultures, it was a precious drink, reserved for people of stature. Montezuma would consume it before sex to give him energy and stamina (kinda puts that Whitman's Sampler you got your Aunt Becky last Valentine's Day in a weird light, huh?)

People considered it a strengthening drink, full of healthy properties, and of such raw power that it was unsuitable for women and children. Sure, drinking too much could make you deranged, but in moderation, it was thought to be invigorating. You know, like Gatorade.

Chocolate wasn't just considered to be an aphrodisiac, not by a long shot. It was also used as a treatment for angina, dysentery, dental problems, indigestion, constipation, fatigue, hemorrhoids, and kidney disease.

Christopher Columbus encountered cacao beans during one of his trips to the Americas, although he referred to them as "almonds," in case you need another reason to think that cat was a total ding-dong.

Spanish conquistador Hernán Cortés might have been the first European to encounter chocolate itself, but it's unclear who actually brought chocolate to Europe. Regardless, by the 16th century, it was a hot commodity all across the continent.

FUDGE YOUR PRESCRIPTION

The predominant system of medicine in Europe at that time was the humoral system.

The humoral system was based on the belief that the body contained four titular humors that had to be kept in balance. Have too much of a certain humor? Purge it! Is the old yellow bile tank running a little low? You've gotta top it off! In addition to the four humors (blood, black bile, phlegm, and the aforementioned yellow bile) that you had to keep in balance, there was also a belief that everyone had a personal climate of sorts. Babies were hot and humid, young adults hot and dry, adults cold and dry, old people cold and humid. Various food and drink items also had corresponding temperature and humidity as well. In order to restore balance to your system, you just ate or drank something that's the opposite of whatever you have going on.

In light of this, cocoa was seen as a cool, dry drink to ward off hot and humid conditions. It could also be modified depending on the diagnosis. Alone it could treat the liver, breast issues, or the stomach. Add gum, and it could stop diarrhea.

Okay, wait—stop the book. We've talked about a lot of really horrific things in this book, and I'm sure we're gonna have a few more gems before our time together is through. But can we all agree that none of them will top . . . chocolate gum? Ugh, I just got the yuck chills typing that.

Want more? Just add corn and vanilla to your chocolate to make a paste that was used as an aphrodisiac (we're assuming it wasn't applied topically). In addition, chocolate was added to a variety of plants to help skinny people gain some weight. (Note: This would probably work even without the additional plants.)

USE AS DIRECTED

This view of chocolate as medicine was especially important because European religious groups started denouncing it almost immediately.

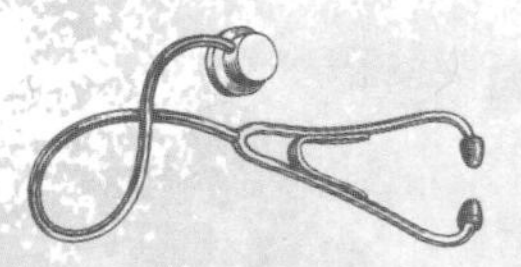

Sydnee's Fun Medical Facts

One particularly popular recipe to fix people who were not feeling well, but did not have a fever, was to add sugar, cinnamon, vanilla, cloves, anise, and chili powder to cocoa and serve as a warm drink. The popularity of this cure is not hard to understand. Did I mention you could also add almonds if you wanted to?

Chocolate was seen as a substance that could invigorate you and encourage sinful impulses, so the only acceptable reason to use it was if a doctor told you to. In response to this, doctors cleverly expanded its uses to basically everything. It was prescribed as a diuretic and an expectorant, and in a particularly strange and unappetizing recipe, cocoa could be mixed with ground human skull, musk, and ambergris to treat . . . hypochondria. There's a pretty good chance that one worked, though, right? We'd guess that after one nibble of that little concoction you'd probably never admit to being sick again.

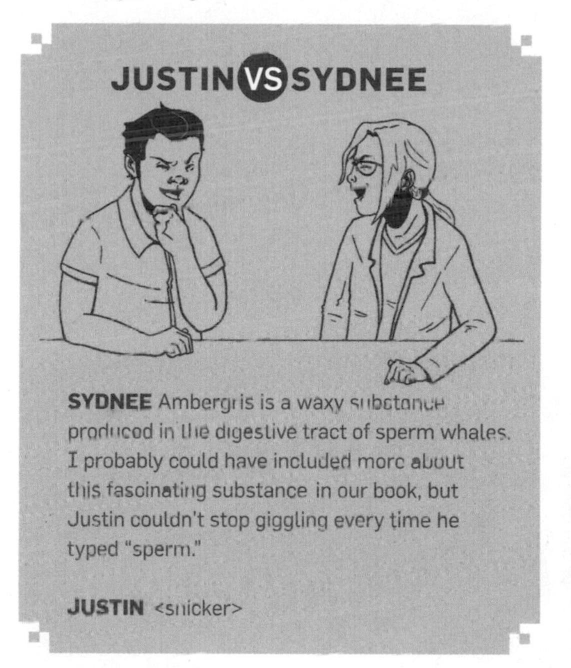

Word of chocolate's miraculous healing properties spread across Europe, particularly as a delicious way for the sickly to gain weight, to stimulate the nervous system, and to improve digestion.

In the 16th century, Grand Duke Ferdinando I de' Medici claimed chocolate was great for flatulence.

By the 17th century, its use had spread even further, and for more conditions. As one 17th century French doctor wrote:

"The use of chocolate is salubrious [for] it excites and strengthens with its warm mild juiciness the bowel's inborn warmth and strength, it helps digestion, it fosters the spread of food and the secretion of the unnecessary, it accumulates fat, it is not an enemy to the brain, it is Venus's friend and very suitable for body and soul."

The preponderance of chocolate being used medically at last led to a showdown over the therapeutic viability of the sweet in Florence, by time of the 18th century.

THE GREAT CHOCOLATE BATTLE

In the early 1700s, Doctor Giovan Battista Felici (who later became known as the Great Chocolate Accuser) stirred up all kinds of trouble, as his nickname might suggest. Felici was certain that cocoa was mislabeled as a cold substance, and was especially hot when you added all the yummy spices to it, as was the fashion.

According to him, this misuse of chocolate could ferment your blood, and it would spoil . . . your . . . blood? It's hard for us to say exactly—get deep enough into humoral medicine, and you're basically playing Calvinball.

A brief diversion, if you'll allow it.

If you're anything like me, you probably aren't familiar with the Grand Duke Ferdinando I de' Medici. We trot out trivia about people with fancy titles like that, and it sounds like they were an impressive person. But you've probably never read that name until about thirty seconds ago right? No, the only reason you're (perhaps only fleetingly) aware of Ferdinando I de' Medici, fifth son of Cosimo I de' Medici and Eleanor of Toledo, is that one time he said chocolate helps you not fart so much.

That's all by way of me saying, you know, it's all really just anybody's guess, huh?

Chocolate sellers were worried that this negative view of their product by a learned doctor, no less, could be bad for business. One guy, Francesco Zetti (uncharitably known by his contemporaries as "the Hunchback of Panone" due to a back condition), decided to take action. He commissioned a report in defense of chocolate from an anonymous doctor and published it. This report got a lot of backing from local chocolate producers, and he ended up winning the battle of public opinion. It probably didn't hurt that chocolate is, of course, delicious.

By the latter years of the 18th century, the war was over, and chocolate was well established as a medicinal beverage, usually served melted and added to milk as a restorative or invigorating drink. It was in such demand that it was suggested that if you couldn't afford chocolate, you could make a fake by toasting flour and mixing it with sugar and milk.

I don't know how this tastes, but for the record, I think I saw a participant on *Worst Cooks in America* try it, and the judges didn't seem to enjoy it. Seriously, Google it. It happened.

By the 19th century, chocolate had become such a popular health beverage that a British Quaker named John Cadbury started promoting the practice of drinking chocolate as an antidote to drinking booze. While Cadbury might have initially been trying to bolster temperance, which he thought could fix many of society's ills, it ended up being a pretty good financial decision too. You might have seen his name on a Creme Egg if you're in the United States, or if you live in the United Kingdom, or most any place chocolate is sold.

Chocolate was also considered a healthier alternative to tea, fattening up sickly kids stuck in factories and providing one with energy, vigor, and robustness. This effect wasn't just important for children. According to one testimonial, a husband using chocolate to treat a respiratory condition shared a little with his wife, and she was able to get pregnant despite being thought to be infertile.

So at this point in history, chocolate has become just about the most symbolically loaded and perplexing gift you could get for somebody, right? Are you telling your sweetie you think they drink too much? That they need to fart less? That they need to gain weight? That you think they should get pregnant despite medical professionals agreeing that it's unlikely to occur? Gifts shouldn't need to come with fifteen-minute disclaimers; honestly, just go with a nice dish towel. Everybody always needs another one.

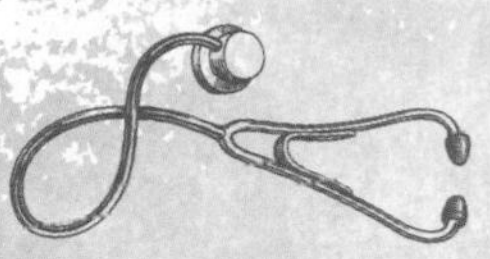

Sydnee's Fun Medical Facts

Hershey's did make one product meant only for nutritional purposes. The "D-ration" was a bar created just for the military in the late 1930s. The demands Uncle Sam put on Hershey was that the new bar had to be nutritionally dense (like over 600 calories) and taste "a little better than a boiled potato" but not so good that soldiers would eat it for fun.

Unfortunately, the beautiful lie of chocolate as health food began to die out in the 20th century with the revelation that too much fat and sugar are actually bad for you. This staggering concept was well established by the 1950s, and chocolate was simply marketed for its most obvious advantage: its yumminess.

MELT IN YOUR MOUTH, NOT IN YOUR HAND

The only thing holding chocolate back was really the method of consumption. Chocolate wasn't shelf-stable, so it couldn't be stored for long times without going bad. And melting it into milk was kind of cumbersome and time consuming. This all started to change in 1828, when Casparas van Houten, a Dutch chocolate-maker, invented a way to remove fat from cocoa beans. His son Coenraad went one step further, creating a powder that had a sweeter taste and mixed easily with water. Known as "Dutch-process chocolate," this powder was easier to store and work with. Of course, new forms of chocolate meant new ways of turning it into medicine. "Health chocolates" like "Dr. Day's Chocolate Tonic Laxative" and "Hauswaldt Vigor Chocolate" began to pop up in the marketplace. Some modern products, like laxative Ex-Lax or calcium supplement Viactiv, have carried on the tradition.

Even chocolate products that were sold for enjoyment and not medicinal benefits were billed as being good for you. The Heath bar was advertised as a health food, using the slogan "Heath for Better Health" because it was made from high-quality ingredients. (Ah the '30s, when any product that didn't actively poison consumers got to be a health food.) Milton Hershey also came around this time and changed the face of chocolate in the United States. The old Hershey's Chocolate ads also used the health benefits as a selling point.

WHAT HAVE WE LEARNED FROM THIS?

Well, the days of prescribing pure chocolate are, sadly, long gone. While there is some evidence that dark chocolate may be considered somewhat heart healthy because it can improve HDL levels (that's good cholesterol), no reputable physician is suggesting you start taking it nightly with your Lipitor. Most people tend to consume chocolate in less-than-healthy forms that are loaded with sugar, so the risks far outweigh the benefits in terms of medical application. There are just simply safer and more effective ways to keep your arteries clear. Let's be honest: Chocolate is really good, and you should probably just stick with eating it in moderation.

JOHN HARVEY KELLOGG

1852–1943 · USA

The two of us were of . . . differing opinions about how to approach this section on John Harvey Kellogg. In the interest of fairness, we're both going to present our biographies.

CEREAL HERO

This legendary Baron of Bran, this Captain of Crunch, was born in Tyrone, Michigan, to John and Ann Kelloggm and studied at the prestigious-sounding Hygeo-Therapeutic College of New York under Russell Trall (who sounds pretty important himself).

Kellogg cared a lot about nutrition, and even opened a health resort called "the Battle Creek Sanitarium" to help people live a healthier lifestyle. While managing the Sanitarium and conducting pioneering research into clean living, he also never stopped trying to create with delicious ways for people to eat healthier.

His most amazing creation, however, actually began with an accident. His younger brother, Will Keith Kellogg, left some cooked wheat unattended while he and John worked on some Sanitarium business. When they returned, they found that the wheat dried out.

The Kelloggs ran a tight ship, though, and rather than let the wheat go to waste, they pressed it flat and baked it, turning it into flakes. The cereal was a hit with the patients, and John and Will started experimenting with other grains. It wasn't too long until Corn Flakes were born.

It was also Will Keith Kellogg that took the idea and ran with it, adding sugar to the flakes and mass-marketing them with the Battle Creek Toasted Corn Flake Company.

I'll admit that I've been trying to avoid some of the darker parts of the Kelloggs story, but there's just no way to stave it off any longer. It's true: Will's decision to set out on his own (not to mention adding sugar to their healthy cereal) hurt John deeply, and it drove a wedge between them that lasted for decades until the elder Kellogg's death.

With that dark segment now fully drawn into the light, we can now move on to celebrating how far the invention of Corn Flakes has taken the Kellogg name. Based on the foundation of that one incredible cereal, the company Will founded, which we now today as Kellogg's, has grown into a multi-billion-dollar corporation. Would we have Crispix, Apple Jacks, or Marshmallow Alien Berry Froot Loops without John Harvey Kellogg's amazing invention? The question is frankly too terrible for us to contemplate.

John Harvey Kellogg, I think I speak for my wife and I both when I say simply, *thank you.*

ANTI-SEX ENEMA FANATIC WHO CAGED GENITALS

While John Harvey Kellogg did indeed study medicine at the Hygeo-Therapeutic College of New York under Russell Trall, I think it is important to also make note of the fact that this particular institution was not your typical medical school. In terms of managing disease, they advised a vegetarian lifestyle and exercise, which sounds harmless enough, I guess. But they also recommended the avoidance of all medications.

It is true that the Battle Creek Sanitarium promoted aspects of a healthy lifestyle, such as a grain-based vegetarian diet, exercise, fresh air, sunshine, and good posture. To make things a bit harder, they also advised avoiding spices, condiments, alcohol, tobacco, caffeine, and sugar. Fair enough—not a lot of health benefits in these particular substances. However, in order to really understand this program, it's important to discuss Dr. Kellogg's beliefs on the origins of disease, specifically that it all either came from the bowels or from sexual intercourse. He advised daily enemas (especially yogurt ones) to keep the bowels clean, and complete avoidance of sex in order to maintain health. Many of the off-limits foods were banned because they were thought to be aphrodisiacs or sexual stimulants.

The origin of the corn flake was really the result of searching for a very bland food that wouldn't get anyone all hot and bothered. He was also not afraid to practice what he preached, claiming that he and his wife of forty years never had sex.

While the whole lifelong feud between brothers thing is indeed sad, I would hesitate in calling it the darkest side of the Kellogg family legacy. Dr. Kellogg's promotion of abstinence was much more extreme than most. He also strictly forbade masturbation, claiming that it caused "cancer of the womb, urinary diseases, nocturnal emissions, impotence, epilepsy, insanity, mental and physical debility, dimness of vision, and moral corruption." While masturbation was also supposedly responsible for cases of paralysis and club foot, he had a solution for that: circumcision.

This was not the kind of circumcision you may be familiar with today. Kellogg actually advised against infant circumcision. He counseled that it should be done without anesthesia and at an age at which they can remember it, to create a negative association with their penises and deter kids from touching themselves.

To be very certain that these kids didn't explore that region of their body anyway, Kellogg offered a variety of other solutions. You could simply watch the all the time so they can't, try bandaging or tying their hands, covering their genitals with patented cages, sewing the foreskin shut, or applying electrical shocks to the area if they touch it. This was just for patients with penises though. For those with a clitoris, he recommended that the parent or doctor apply carbolic acid to the clitoris to prevent masturbation by blistering it, or simply remove the clitoris altogether. A surgery that he was willing to perform.

I hate this guy.

If that hasn't swayed you from the vision of John Harvey Kellogg as simply a kindly old cereal genius, I should probably also mention that he believed in eugenics and the separation of the races. The only good thing I can say for the Great Depression is that it ended the popularity of the Battle Creek Sanitarium, and though Kellogg tried to open a new one in Miami in 1931, it never really caught on like the first one did.

That is the true and very dark origin story of Corn Flakes.

Also, they don't taste very good.

Parrot Fever

This is not a chapter about Jimmy Buffett. We know, fellow Parrot Heads, we're disappointed too.

As much as we'd love to spend 2,000 or so words taking a tour of the Mayor of Margaritaville's many musical masterpieces, our boring publisher shut us down. Their total snore of an argument was that being really into Jimmy Buffett isn't technically a medical condition, and they're the ones signing the checks.

So if you want a chapter on Parrotheaditis, you better buy a lot of copies of this book so we'll have the pull to throw our weight around and demand its inclusion in the sequel. For now, you'll have to settle with us exploring the hidden dark side of these beautiful tropical birds.

You might be tempted to believe every parrot you meet is your friend, and wouldn't hurt a single hair on your beautiful head. That's especially true if every time you see a parrot you say, "I'm a pretty bird who's your friend, and wouldn't hurt a single hair on your beautiful head," and wait for it to repeat it back to you.

Don't believe their lies.

CATCH THE FEVER

Now, the scientific name for this disease is not actually "parrot fever." It is, in fact, psittacosis, derived from the Greek word *psittakos*—which, of course, means "parrot." . . . Okay, disregard that first sentence. It is actually named parrot fever; it's just in Greek so it sounds more impressive.

Psittacosis is a bacterial infection that can manifest in a variety of ways in humans. The most likely presentation would be fevers, chills, headaches, and coughing. However, it can result in more serious infections, including pneumonia, or a liver or heart-valve infection.

The disease clambers into your body in what is one of the most unpleasant and unlikely ways possible. The bacteria is generally spread by inhaling dried bird droppings. The Centers for Disease Control also notes that the infection can be transmitted by direct "beak-to-mouth contact," although, thankfully, this is thought to be fairly uncommon.

While I'm happy that means that fewer people are getting parrot fever, it does make me a little sad for our nation's feathered companions. Come on, y'all. Polly has plenty of crackers. You know what Polly really wants? A deep, passionate kiss from your mouth that lasts for a length of time outsiders would describe as "deeply unsettling."

While parrots get all the blame for this one, the disease can actually infect many pet birds including parakeets, macaws, and cockatiels, as well as chickens and turkeys. It is a fairly rare condition and, luckily, treatable. It is fatal in less than one percent of patients, as long as they receive appropriate treatment.

The first descriptions of a similar illness date back to the late 1800s; and it was officially given its avian moniker in 1895. Usually we'd dwell on all the ways society has tried to treat Parrot Fever, but we didn't actually isolate the bacteria that was causing all the trouble until an epidemic in the late 1920s. So let's start our story there.

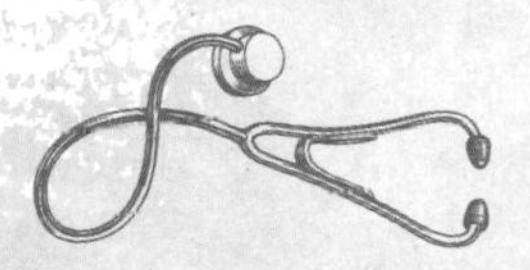

Sydnee's Fun Medical Fact

Okay, so this isn't so much a fun fact as it is an admission. A lot of big scientific-sounding words are really just the Latin or Greek for regular words. Supposedly, we doctors use them so that we can all share a common scientific language. I suspect we also like how much more impressive it seems to diagnose someone with "otitis media" than "earache," but maybe that's just me.

BIRDEMIC BEGINS

In 1929, a large shipment of Brazilian birds arrived in Argentina to be sold to, well, the sort of person who would have wanted to buy a Brazilian bird, we guess. You know the type. The problem was that these birds were visibly sick: Little hot-water bottles on their heads, thermometers dangling out of their mouths, the whole nine yards. The dealer, in the grand tradition of all carnival barkers offering goldfish as prizes, decided to keep their precarious health to himself.

The unscrupulous bird salesman unloaded the sickly birds, and it wasn't too much later that reports of their new owners taking ill started popping up. The most notable was an Argentinian actor who played a sailor, and used a live parrot as part of his performance. Did the ailing avian hit its marks and remember all its lines despite adversity? History has sadly let this information slip into the ever-shifting sands. No, the important thing is that that unlucky buyers in twelve different countries began to fall ill after purchasing the pallid parrots.

In the United States, Simon Martin, secretary of the Annapolis Chamber of Commerce, bought a parrot for his wife for Christmas. We'll assume she actually wanted one, perhaps dropped a few hints, because otherwise that is one heck of a long shot in the spousal gift department. In order to maintain the surprise, he asked his daughter and son-in-law to keep the bird at their place until the big day.

Like, can you even imagine? A bird in your home? I know I joked about smooching birds earlier, but come on, a real bird that lives in your house? I'm getting the chills just thinking about it. If my dad shows up with a bird in his passenger seat for me to care for, his backseat better contain a kidney that I desperately need for my continued survival because, otherwise, no way.

Unfortunately, by the time Christmas had arrived and the present was to be delivered, the parrot had passed away. Let's hope that this was discovered before the big reveal on the twenty-fifth, the alternative is just too bleak to consider.

As if the parrot's passing wasn't enough to dampen this poor family's yuletide spirit, the daughter and husband who were nice enough to keep the bird had also fallen ill.

Do you see? Do you see now, reader? You thought I was overreacting before, but this is what happens!

PANIC TAKES WING

When their family doctor heard about the strange symptoms and the dead bird, he was reminded of some cases in Argentina that he had recently read about—specifically the sailor impersonator who apparently had been high profile enough to make headlines. After he put together the pieces, he alerted the U.S. Public Health Service that there may be some sort of strange epidemic developing. Through Martin, the mayor of Baltimore learned of the outbreak, and from him, the governor of Maryland was alerted. As the unsettling game of telephone continued, panic began to spread.

City and state health departments sprang into action, working tirelessly to understand and hopefully defuse the public-health crises. They were joined by the National Health Service and representatives from the Army and the Navy, all of whom focused their efforts on Annapolis. At this point, the pet shop that had sold the parrot had started receiving more calls about sick and dying birds. While that made it clear that the illness was in some way connected to birds, that's about where certainty flew the coop. In grand human tradition, the best advice authorities could cook up? An all-out, unfounded, relatively one-sided, no-holds-barred war on parrots.

Sailors were told to throw their parrots into

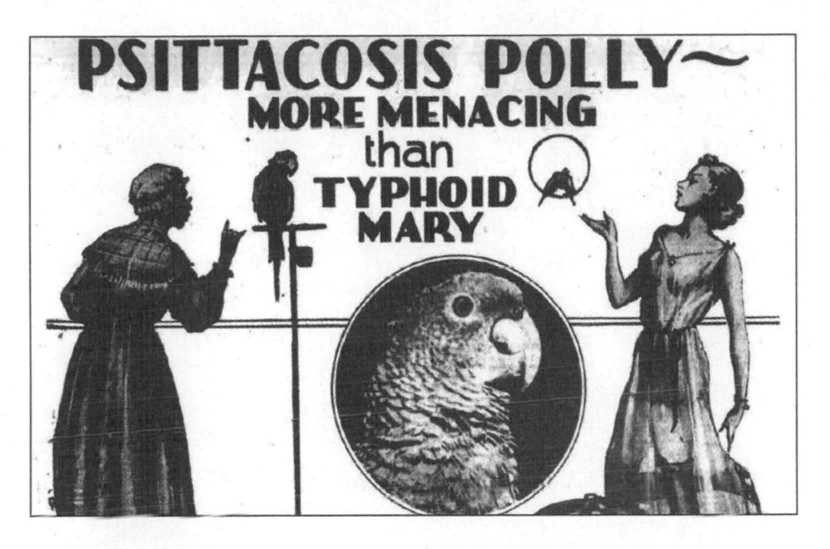

the sea. Some pet owners just began setting their parrots free even though they were advised by the authorities to choke their birds to death.

It's just like *Old Yeller*, but with a fun tropical spin . . . and also if the kid had to strangle Old Yeller to death with his bare hands—okay, so it's not really so much like *Old Yeller*.

As often happens when the public begins to panic and stops thinking clearly, misinformation and gossip spread very quickly. In Toledo, it was reported that an elderly woman had died of pneumonia just days after her husband brought home two parrots from Cuba. In Baltimore, another elderly woman died of pneumonia, and while she didn't actually have any birds, everyone was fairly certain that she had recently touched a parrot somewhere. This kind of panicked reporting lead to many false alarms, necessitating evening news reports that included updates on parrots found to be healthy.

That may seem laughable, but we think we speak for all reading when we say it'd be a welcome relief to turn on the news and see "HERE ARE SOME COOL HEALTHY PARROTS WE HEARD ABOUT" scrolling across the bottom of the screen.

When pet birds passed away, many people began boxing them up and shipping them to Washington in an effort to be helpful. But who

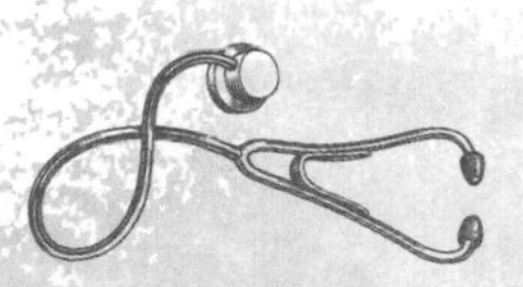

Sydnee's Fun Medical Facts

How do you tell if a parrot has parrot fever? Well, I'm a human doctor, not a vet, so I didn't know, but I did go look it up. Apparently sick parrots will have puffy eyes, lethargy, loss of appetite, fluffed feathers, a runny nose, and an enlarged liver. I don't know how a bird owner figures out that last one, but there you go.

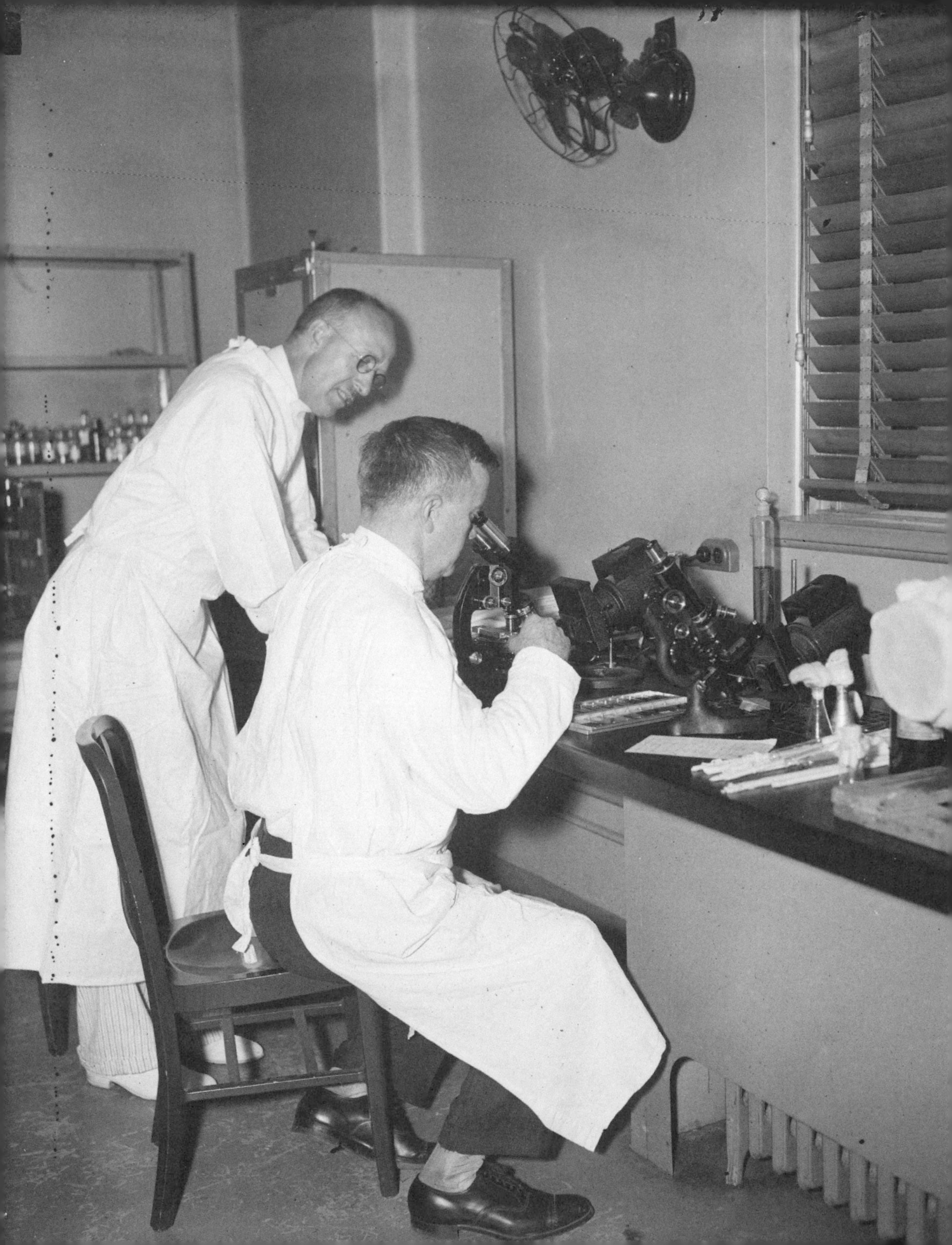

was the lucky person on the receiving end of all these packaged, deceased Christmas parrots?

THE BIRD BRAINS

The Hygienic Laboratory in Washington, D.C., had placed Dr. Charles Armstrong in charge of investigating all the samples and identifying the bacterial culprit. It is worth noting that the Hygienic Laboratory was hygienic in name only. The researchers there were of the mindset that you didn't want to waste time being too careful because then you might not get the answers you need in time to help people. Also, someone else might beat you to the solution.

There weren't a lot of government dollars floating around to support scientific research at this point in history. The nation was still reeling from the stock-market crash, and money was tight. In order to claim your share of the pie, you had to have exciting results—and you had to have them fast. All the media attention had put parrot fever in the national spotlight, and made it a prime target for those seeking government grants.

Unfortunately, then as now, public attention can be short lived. Even as employees at the pet shop were beginning to fall ill, the media began to lose interest. It turns out that a lot of the early news reports were either false or overblown, and the whole thing started to become more of a joke than an actual scare. This was not great timing, considering that the researchers at the laboratory (including Armstrong) were starting to get sick as well. Lots of them. Some of the researchers actually succumbed to the illness.

In the midst of chaos, Dr. George McCoy, director of the lab, decided to take action. He assumed control of all the research, but that was just the beginning. Dr. Armstrong was still fighting to survive in the hospital at this point, but McCoy wasn't content leaving his fate up to chance. He took some blood from a patient that survived parrot fever and went and injected it into Armstrong.

It's worth noting that no one was doing this at the time. But it's even more worth noting that no one is doing this now. Don't do this. Yes, Dr. Armstrong did survive, but there's no way of saying if McCoy's weird Hail Mary was actually effective.

Unsatisfied with simply (possibly) saving a man's life and faced with a fifth of his workforce out sick, McCoy made an even bolder move. He evacuated the building and then marched in alone armed with chloroform. He then killed every animal in the lab, incinerated their bodies, and sealed the lab. He called a fumigation squad to finish the job, and they sprayed the building with cyanide. It was such a thorough job that completely healthy, probably really polite sparrows who weren't bothering anybody dropped out of the air dead overhead.

IS THIS STILL A PROBLEM TODAY?

By the end of the epidemic, 169 people contracted Parrot Fever and thirty-three died. This number seems alarmingly high, and, indeed, any loss of life is a tragedy, but remember this was before the age of antibiotics. These days, Parrot Fever is very treatable and rarely fatal.

If anything good came out of this tale, it was what became of the Hygienic Laboratory. In the wake of this drama, representatives from the lab took their complaints to Congress. The lab had been poorly and recklessly run, and it was by McCoy's extreme tactics that more employees hadn't gotten sick. Hygienic Lab workers argued that the Parrot Fever debacle was a prime illustration of why the institution needed more funding and more power. Two months after the Parrot Fever epidemic, Congress acquiesced; the Hygienic Laboratory was given everything they asked for, including a much better name.

Today we know it as the National Institute of Health.

So What's the Deal With:

DETOXIFICATION?

Because we don't have enough real medical problems to worry about, let's say we invent one!

So, How Long Has This Been a Thing?

Normally our purview is medicine's hilarious past, but let's pause for a moment to consider a largely depressing, but still a little funny, part of its present. "But Justin and Sydnee," we hear you saying, "detoxing is a really old idea; certainly it has some merit!". Well, for one thing, it's very rude to talk to a book. Secondly, go Google "Appeal to Ancient Wisdom." Folks, if you gain nothing else from *Sawbones*, we hope you'll remember that just because people keep doing a thing—even if they do it for centuries—that doesn't necessarily mean that it works.

Anyway, the ideas behind detoxification in ancient times were derived from humorism, the debunked medical theory that your body contains four vital fluids which must be kept in balance to maintain health. If you build up an excess of one of these fluids, or humors, you can regain equilibrium by expelling it through one or another of your bodily orifices.

Hippocrates particularly believed in fasting to purify the body, mind, and soul. Some other traditional purification rites such as sweat lodges, and even saunas, have been seen as perhaps equally cleansing to body and spirit.

The Indian Ayurvedic tradition has a detailed and specific detoxification system called the *panchakarma*, or five treatments. It calls for a rigorous regimen of vomiting, enemas, laxatives, and/or bloodletting, followed by a neutralization phase with herbs, heat, and warm foods to balance you back out, once you've vigorously expelled all those toxins.

So, Does This Work?

Getting rid of the toxins in your body is a good and noble impulse. But we have good news: Your body is already doing it. In the medical sense, detoxification is the process of your body slowly clearing itself of a toxin, often a harmful drug, through physiologic processes (which is to say peeing and pooping). Your kidneys and liver are well suited to this task, and the detox regimens advertised are generally worthless and occasionally downright dangerous.

A lot of these extreme fasts and purifications can leave you dehydrated, malnourished, and kind of gross-smelling. Or maybe even anemic if you fell for the brief, but bloody, leech-based Hollywood health craze. Don't remember that?

Well, they were indeed a thing. Demi Moore promoted them on David Letterman, saying "they have a little enzyme that when they're biting down on you, gets released into your blood and generally you bleed for quite a bit. And your health is optimized. It detoxified the blood, and I'm feeling detoxified right now."

This is a pretty good moment to remind you that if someone, anyone, credits anything other than bodily organs with detoxification, it's probably a good time to leave the room . . . or distract them by asking what it was like to be married to Ashton Kutcher.

What's the Deal?

The thing is that our bodies are really good at dealing with toxins. It makes sense that there's heightened interest in detoxification, given that we perceive our environment as being more toxic that it once was, and we hear so much about the chemicals and other pollutants we're probably ingesting. So, we wanna get all scrubby clean inside and out.

In reality though, unless you've actually been poisoned, your body is doing all the detoxing you need. You definitely don't need to go to such extremes as coffee enemas, pulling out your fillings (really, don't do that), or vigorously brushing your skin with a dry brush morning and evening. Though if you're going to pick one, go for the dry brush. At least you'll be exfoliated.

MIRACULOUS UNIVERSAL CURE-ALL

VINEGAR

Though it's currently having its moment in Gwyneth Paltrow's Goop-induced spotlight, vinegar has been used for health applications for literally thousands of years . . . for *everything*. No, seriously, everything. You'll see in a minute.

You know what vinegar is, of course, but have you ever thought about how it's made? No? Well, you've already started the paragraph, so let's just get through this thing. Vinegar (which comes from the French for "sour wine") is made via the fermentation of ethanol alcohol. Bacteria breaks the alcohol down into different components including acetic acid, which gives vinegar its unique character. That's a very scientific way of saying vinegar is a weird, thin, stinky soup that gives Easter eggs their crazy color. Or something.

At the risk of veering into TMI territory with vinegar, we'd be remiss in ignoring the culture that creates the vinegar, which is called the "mother," and congeals into a gross slime at the top of the vinegar unless it's pasteurized. Some health-food enthusiasts encourage consumption of the mother of vinegar mother, but the science behind that is . . . less than impressive. It's got a lot of iron apparently, though.

Okay, we've hyped you up on this gnarly juice enough, it's time to list a bunch of things you probably shouldn't use it to treat even if an old guy assures you it'll work.

USED TO TREAT:

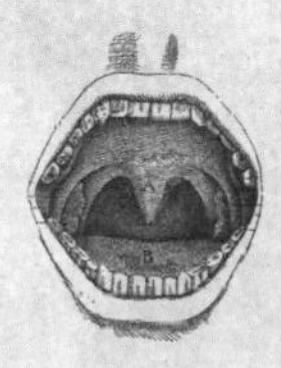

SORE THROATS

Ancient Greeks were big on oxymel, a blend of four parts honey to one part of any vinegar that was then simmered down. The finished concoction would be kept on hand in order to treat all manner of ailments, but sore throats were a popular target. Oxymel is still commonly used by herbalists today who infuse it with (no points for guessing) herbs.

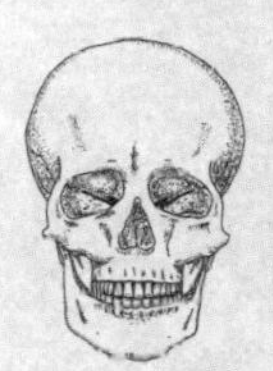

HEAD WOUNDS

Hippocrates liked to use vinegar for a variety of maladies (sensing a theme?) including ulcers and difficulty breathing, but our favorite was head wounds. A word of warning: It was better for those with too much yellow bile than those with too much black bile (see page 110 for more tips on keeping your humors in balance). Oh, also: It's better for men because it could irritate a woman's uterus.

CORPSE CLEANLINESS

Song Ci, considered the father of forensic medical science in China, used his book *Collected Cases of Injustice Rectified* to advocate for washing hands with a combination of vinegar and sulfur when handling corpses.

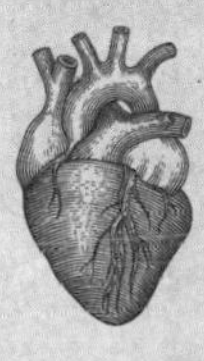

HEARTBURN

This is one of those persistent folk remedies you see quoted everywhere, but rest assured, vinegar has not been proven to have any effect on heartburn. Heartburn, by the way, is caused by acid trickling into your esophagus from your stomach, so an acid—like vinegar—might even make it worse.

SCURVY

As far back as the American Revolution, soldiers received rations of vinegar to help prevent scurvy. They even tried this on the Lewis and Clark expedition. You could understand the logical leap that was being made (vinegar is tangy, like citrus fruits), but with no vitamin C, the vinegar wouldn't have been much help.

WINNING BETS

Okay, this one isn't medical, but Cleopatra once bet Mark Antony she could eat a meal worth 10,000,000 sesterces ($500,000 these days). She won the bet by dissolving a pearl in vinegar (Pliny said it was "the largest [pearl] in the whole of history" but who knows with that guy) and drinking it. It sounds like an apocryphal story, but it's all theoretically possible. Try it yourself with *your* giant pearls!

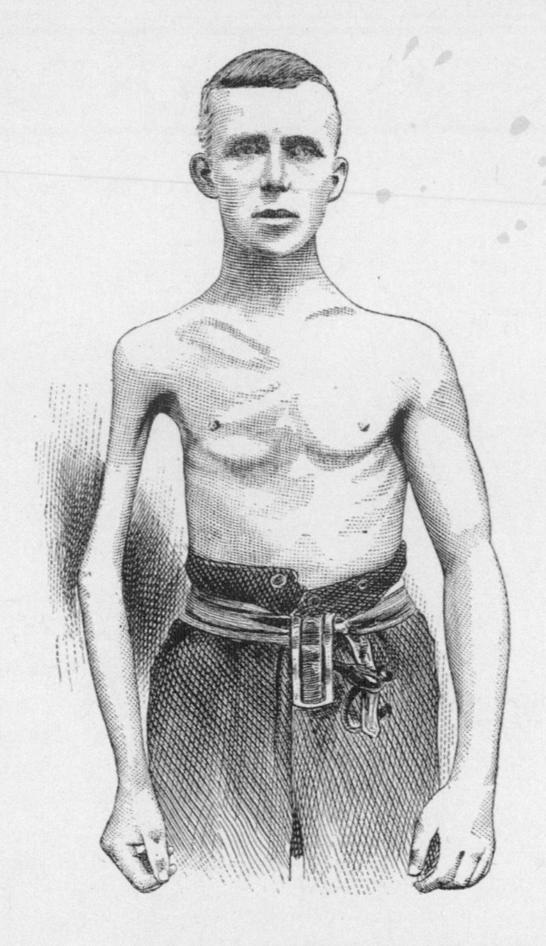

Polio Vaccine

One time, we got it right.

There are so many stories in this book about quacks and bad medicine that it's nice to sometimes consider the times when we got it right. The story of polio treatment is a story with some real heroes in it. Also, since half of the authors of this book are doctors, it is nice when a story isn't about how dumb or evil they are.

Polio is a virus, classified as an enterovirus, and it is usually fairly benign in the way it interacts with the human body: It enters through what is known as the fecal-oral route.

Wait, I'm no Latin expert, but am I correct in assuming that means . . .

Yes, that's exactly what it sounds like.

Polio takes a while to incubate, up to twenty days in some cases, and surprisingly, may not result in much at all. As scary as polio may be, 95 percent of those who contract the disease are symptom-free and don't even know they have it. Unfortunately, though, they can still pass it on to other people. In about 4 percent of cases, you get some mild symptoms, maybe a cough or runny nose, or perhaps some abdominal pain and nausea, or maybe just body aches and malaise. The scary part is that in about one percent of cases, the virus invades the central nervous system, and this is when it can result in paralysis. While those may seem like good odds, the polio virus is highly infectious, which means a whole lot of folks are exposed during an outbreak, and well, it means a lot of very sick people.

Plus, I mean, one percent isn't that low, right? Like, if I heard there was a one-in-one-hundred chance to hit the Powerball, I'd hock everything to buy more tickets. I'd sell my kids on the black market. Well, one of my kids. . . . Okay, none of my kids, but I'd sell my Nonnee on the grandma black market, and I *love* my Nonnee.

While throughout most of medical history we can credit sanitation for improvements in disease outcomes, polio is an anomaly. Improved sanitation probably actually made things a little worse in terms of the severity of polio. Initially, most cases were among infants who actually had a lower chance of suffering long-term effects from contracting polio. The rate of paralysis increases significantly as you age. Once we understood the importance of sanitation, babies were among the first to benefit from less exposure to disease. But fewer people contracting polio when they were an infant meant the average age of incidence rose from between six months and four years, to between five and nine years old. There were also many more cases of teens and adults contracting polio, which tragically meant many more cases of paralysis and death.

We're getting a little ahead of ourselves, though. Polio is a disease that goes all the way back to the beginning, with recorded cases throughout antiquity. There are paintings and descriptions from ancient Egypt that depict a type of paralysis that occurred in children. While we can't be sure these images of children with canes reflected polio, it's very likely. It is also likely that Roman Emperor Claudius had polio. Sir Walter Scott got polio in 1773, and lost the use of one leg as a result. In his account of the illness, he refers

to it as a "severe teething fever." Polio may have also crossed over into fiction: Many believe polio plagued the beloved, doomed Tiny Tim before Scrooge saved the day. Again, it's impossible for us to say for sure.

For centuries, polio outbreaks only popped up sporadically. It was known by many names, it was unpredictable, and its cause was unknown. We weren't forced to answer many questions about the mysterious illness until the 1900s when more frequent epidemics robbed us of the luxury of ignoring polio. Doctors began seeing cases with more and more frequency throughout the United States and Europe, culminating—at least, in North America—during the summer of 1916.

THE DEEP END

Kids in 1916 spent their summer vacation time much like kids today. They stood outside stores. They played . . . well, obviously not the PlayStation 4 in 1916; we couldn't have been further along than PlayStation 2 back then. But also, they filled the public pools and beaches, looking for a spot to cool off and relax with their friends. However, that summer, all this would change. What had started with scattered cases throughout several major cities was now turning into an epidemic. There were 27,000 cases of Polio diagnosed that summer, and 6,000 of them proved the be fatal. New York City was hit especially hard, with a full third of those deaths occurring within the city limits.

As the illness spread and number of affected families climbed, people began to scatter from the city to surrounding rural areas. Public places like pools, amusement parks, and beaches were closed. Terrified parents avoided water fountains. The names and addresses of confirmed cases were published in the paper, and those who'd been affected were quarantined. Families who had been stricken by polio put up signs in their windows warding off visitors. It was the first of a few dozen years where summer meant parents living in fear of polio.

Without a firm understanding of what was happening or why, desperate patients tried all manner of strange treatments that were rumored to work. Poultices with mustard, slippery elm, and chamomile were frequently employed, as well as baths in almond meal. Doctors didn't have much better advice. They handed out prescriptions for quinine, caffeine, gold, and water that had been exposed to radiation and was thought to contain healing powers. In order to reverse the paralysis, some attempted applying electricity to the legs. Then as now, vitamin C was recommended when doctors ran out of ideas.

Boy, I don't care if you're talking about Tylenol or spring water imbued with ten parts per million of mountain daisy; there's just nothing that sounds like real medicine when you describe it as having "healing powers."

As epidemics continued every summer, new therapies were tried. President Franklin Delano Roosevelt, having recovered from polio as a child, brought attention to hydrotherapy as a treatment. Hydrotherapy had been around for many centuries already at this point and encompassed a variety of water-based treatments including mineral spas, warm and cold compresses, whirlpool baths, and sometimes, just taking a bath. The use of hydrotherapy for polio largely consisted of physical therapy performed in water, the use of various pressures and temperatures, and bathing in mineral springs. FDR himself had used hydrotherapy and credited it with his impressive recovery. To expand the availability of this therapy, he bought Warm Springs in Georgia, turning it into a polio rehabilitation facility.

DARKEST BEFORE THE DAWN

For nearly forty years, the polio problem continued on a dangerous trajectory, with

increasingly frequent epidemics. 1952 would see the largest outbreak in U.S. history. There were over 57,000 cases, over 3,000 deaths, and over 21,000 people left paralyzed. At this point, surgeons were going to even greater lengths, attempting procedures like tendon and nerve grafting and limb lengthening to reverse the effects of polio. When surgery wasn't enough, polio victims had to rely on newly created and improved devices, including casts, braces, canes, crutches, and wheelchairs to restore their mobility.

One of the key breakthroughs to support those who had suffered some of the most life-threatening effects of the disease was the iron lung. The prototype was initially just an electric motor hooked to two vacuum cleaners. The device would apply positive pressure to make you breath out and then negative pressure to make you breath in. While it would save many lives, the people who were unfortunate enough to have needed it still ended up with a 90-percent mortality rate overall. In modern times, the iron lung was replaced by the ventilators.

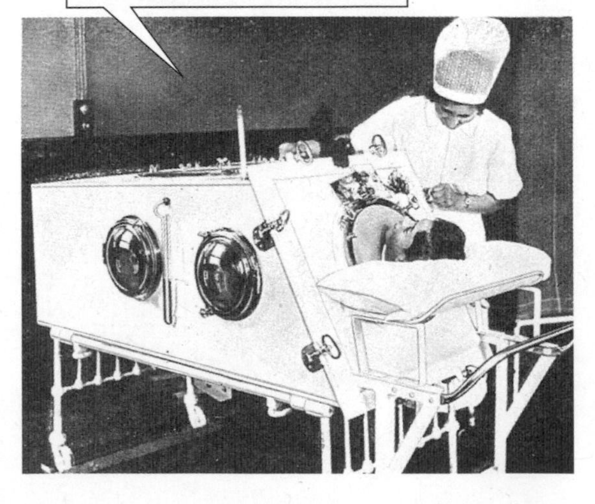

One effective treatment that originated during the 1920s was called the "Kenny Regimen." This therapy was developed by and named for Australian nurse Sister Elizabeth Kenny. (No, she wasn't a nun; "Sister" was the British title for a chief nurse in World War I.) At the time, it was thought that casting patients and keeping them immobile was the best way to prevent contractures. In fact, this only made things worse by causing muscle atrophy. She introduced a regimen of heat, early physical therapy, and exercise that was much more effective in maintaining muscle bulk and flexibility, as well as easing pain. It was so revolutionary that it is actually still the basis for therapies used today.

HERE COMES THE SUN

The solution to polio, though, was in finding a way to prevent it from infecting people in the first place with an effective vaccine. Before doctors could really take the fight to polio, they had to discover how to grow the virus in a culture. This feat was achieved by John Enders in 1949 and won him the Nobel prize a few years later. Now that the virus could be pinned down in a lab, scientists could get to work formulating a vaccine to stop it.

By the 1950s, two doctors, Jonas Salk and Albert Sabin, were hard at work trying to solve the puzzle.

Jonas Salk was a virologist who was tasked with finding an inactivated vaccine by the National Foundation for Infantile Paralysis, established by FDR. A killed virus vaccine was thought to be a safer option. This injection would likely prevent the worse outcomes of polio, but would not stop the initial infection from taking hold in the intestines. Meanwhile, Albert Sabin, a physician and first-generation American immigrant, was working in his own lab to develop a live oral vaccine that, while thought to have more risks, would be able to

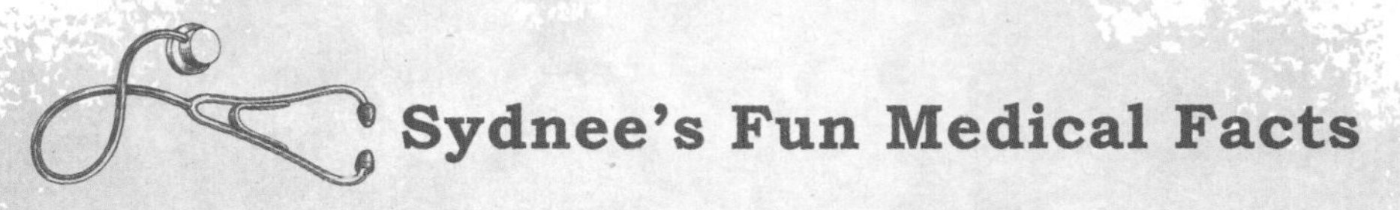

Sydnee's Fun Medical Facts

There are different ways for us to create immunizations. Some involve the use of live viruses that have been attenuated, or inactivated, so that they can no longer infect a person. Others are made with killed viruses or pieces of viruses that will also invoke an immune response in the person receiving them. Both can be safe and effective when made correctly.

prevent the initial infection in the intestines.

Salk was a little bit ahead of Sabin, and was the first to start clinical trials. He was either desperate or confident enough in his cure that himself and his family were among the first injected with the vaccine. After his preliminary tests, he asked for children to test the vaccine on and was met with over a million parents willing to sign their children up for the trials. So great was the fear of polio at the time that parents were willing to risk their children's lives to give them the experimental immunization.

It was nothing short of an astounding success. By 1965, polio cases in the United States had fallen into the double digits, down from over 57,000 just thirteen years earlier. By the 1990s, there were less than ten infections a year—with many years boasting zero cases.

Meanwhile, Sabin had developed an oral immunization that was easier to take and distribute, but was feared by many because it was made with weakened—but live—viruses. Despite successful trials in Ohio, Sabin could not get the oral vaccine the support it needed to have it used nationwide. The Public Health Service was considering its use, but the Salk vaccine was spreading faster with the preferential support of the March of Dimes. The Sabin vaccine did become the standard outside the United States, though, saving millions of lives in the former Soviet Union, Japan, Mexico, Singapore, and parts of eastern Europe.

It's an incredible triumph, but this might be the most astounding thing: Both of these doctors refused to patent their vaccines. They had both found effective solutions for preventing a devastating disease and stood to make millions of dollars by selling the rights. Neither would do so. Sabin never made a dollar off of his vaccine, living only on the salary he received as a professor. Salk even hated the publicity he received surrounding his achievement. When asked in an interview by Edward R. Murrow who owned the patent for his vaccine, he replied, "Well, to the people I would say. There is no patent. Could you patent the sun?"

IS THIS STILL A PROBLEM?

In places where either of the vaccines are in widespread use, polio has essentially been eradicated. The World Health Assembly has had an initiative to completely eliminate the disease since 1988, and they are still making progress. Unfortunately, there are still parts of the world where transmission exists. There is no treatment or cure, so vaccination is still essential as a preventative measure.

But of course, we don't have to tell you how great vaccines are, right?

RIGHT?

THE DOCTOR IS IN

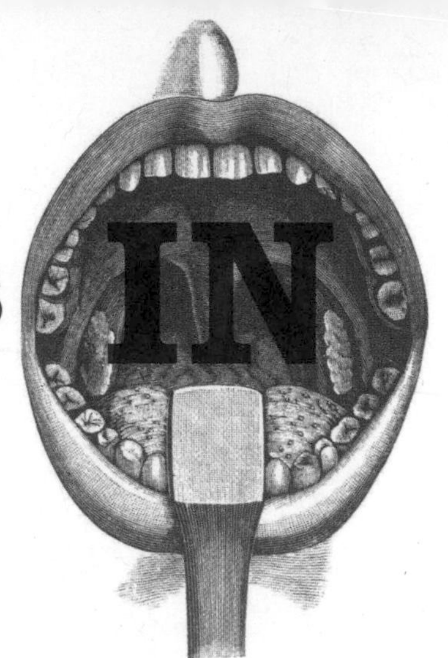

What's that? You still haven't had all your queries about the human body answered? Is that even possible? Well, good news! We're certain that by the end of this final segment, we'll have answered every question about the human body you could possibly have.

Is it possible to get a disease you have been vaccinated for?

Sydnee: While vaccines are a life-saving scientific achievement that prove humankind is capable of amazing things, they are not always 100-percent effective for all people. A vaccine is designed to stimulate your immune system to generate antibodies against a certain disease by exposing it to parts of the virus or bacteria that cause the disease. This is a predictable response in most people. However, there are some whose immune systems will not respond as we would expect. This can be due to immune deficiencies, illnesses, medications, or just genetic differences in the way an immune system responds. We are able to test if a vaccine is working with a blood test called an "antibody titer." If your antibody levels come back too low, you can always consider a booster vaccine to try and get those levels up to where they should be.

Justin: Alternatively, I have perfected a method of boosting my resistance to many common airborne illnesses called "staying on my couch unless I am physically forced to do otherwise." It's hell on my social life, but I haven't sneezed in thirteen years. Well worth it.

Why does chemo cause your hair to fall out?

Sydnee: Not every drug used for chemotherapy will cause this effect, but it can be life-altering for patients that are affected by it. The reason this occurs is intrinsic to they way many chemo drugs are designed to work.

Many agents target fast-growing cells, because cancer cells are fast growing. But that can turn other fast growing cells—like those cells at the roots of your hair—into collateral damage, resulting in hair loss. The cells that line your GI tract are also fast-growing, which is why chemotherapy can cause side effects such as nausea, vomiting, and diarrhea.

While this sounds discouraging, these side effects are usually temporary and worth the risk due to the comparative risk of the cancer itself. This is also why researchers are always hard at work searching for new, better, less-toxic cancer treatments that will save even more lives.

Would a person with *situs inversus* require a donated organ to also come from a situs inversus donor?

Justin: I'd like to take a crack at this one, Syd. And I'd like to begin, question asker, by asking a question right back to you. What when we really come down to it . . . is . . . situs inversus? Hmm, it's a thinker isn't it? Syd, before I monopolize the whole thing, I'd be really interested in hearing your take.

Sydnee: Well, Justin, situs inversus is a condition in which a person's internal organs are basically a mirror-image of most humans' anatomy. The heart, lungs, liver, and digestive organs are all on the opposite side from where they generally exist. On its own, this condition is harmless. The organs still function the same, they just aren't where you would have guessed. This is usually caught incidentally when a patient has a chest x-ray, or some other kind of imaging done for other reasons. It is fairly rare (about 1 in 10,000 people), and it may have some genetic component. But that isn't entirely clear yet.

The question as to how this would affect an organ transplant is really interesting and couldn't really be predicted until it was attempted. There have been very rare cases of transplants, specifically heart transplants, in these patients, and the surgical team did not have to use a donated organ from another situs inversus patient. This did present extra challenges for the surgeons, as they had to reconfigure all of the veins and arteries in terms of attaching the new heart in place. This also involved using grafts from elsewhere in the body. However, this has been achieved and could be with other organs from non-situs-inversus donors.

Does the human body absorb any even remotely substantial amount of water through the skin? For instance, if one were dehydrated, would submerging oneself in water help in any way?

Sydnee: The skin is not a very effective route of water absorption. While it is not completely waterproof, it is fairly water resistant. This is due to the layer on the outside of our skin known as the "stratum corneum." This layer of keratin, dead skin cells, and oils protects our deeper layers of skin as well as regulates our hydration by keeping water in. While a small amount of water can get through this barrier with some prolonged exposure, it is not a significant amount and certainly not enough to combat dehydration. Other substances, like certain medications and toxins that are oil-based, can get through this layer of protection to varying degrees. This allows up to apply some medications as ointments to treat localized conditions, but it also puts our bodies at risk for toxic and occupational exposures through our skin. Overall though, if you are thirsty, have a drink, not a dip.

Justin: This obviously does not apply to Gatorade, which is able to hydrate topically through the power of electrolytes. Especially lemon-lime.

So here's a quick one: What is the adjective form of the word "pus?"

Justin: What a weird question, wouldn't it just be "pu"—Oh, Syd! Please, please tell me it—

Sydnee: Sorry, J-Man. If a wound is full of pus—a thick, white drainage with a high concentration of white blood cells—then the medical term for it is "purulent." This typically indicates infection, as all those white blood cells have arrived at this location in your body and heroically given their little cellular lives in order to protect you from some sort of invader.

Justin: This is, literally, the worst day of my life.

INDEX

C

D

E

INDEX

INDEX

Y
yeast infections, 167
yellow bile humor, 110–112
yellow fever, 158–159, 180
Yersinia pestis bacteria, 35
yogurt, 167

Z

About the Authors

Sydnee Smirl McElroy is a family physician and assistant professor at the Marshall University School of Medicine, as well as co-host of *Still Buffering: A Sister's Guide to Teens Through the Ages*. Justin McElroy is the co-creator of *My Brother, My Brother and Me*, an advice podcast and TV show, and *The Adventure Zone*, a role-playing podcast and graphic novel series. He's a two-time recipient of the Associated Press of Ohio's award for best business writing and a co-founder of Vox Media's gaming site, Polygon. Together Justin and Sydnee co-created *Sawbones: A Marital Tour of Misguided Medicine* and two non-podcast daughters, Charlie and Cooper.

About the Illustrator

Teylor Smirl is a graduate of New York's School of Visual Arts whose work has been featured in comics such as *Amazing Forest* as well as her own series, *Flightless Birds*. She's also co-host of *Still Buffering* with Sydnee and their youngest sister, Rileigh. Find more of her work at http://teylorsmirl.tumblr.com/.

About the Show

Sawbones: A Marital Tour of Misguided Medicine is the planet's most popular medical podcast and has been downloaded over 45 million times since it launched on the Maximum Fun network in 2013. In 2017 it won the Independent Investigations Group Award for the promotion of science and reason in podcasting.

Acknowledgements

Justin would first like to thank his mom and dad for literally everything. He'd also like to thank his brothers, Travis and Griffin, for believing and encouraging him every step of the way, Chris Grant for changing his life over a decade ago, and Jesse Thorn for helping to birth *Sawbones* into existence. Also, thanks to my third grade teacher, Mrs. Williams, for telling me I was capable of writing great things. I hope one day to prove her right.

Sydnee would like to thank her mom for starting to teach her to read when she was six months old, her dad for passing along his unmatched tenacity, and her sisters, Teylor and Rileigh, for inspiring her and making her laugh every Monday. I wouldn't have believed I could do this, or practically anything, without the love and support of my Mama and Papa. Thank you for sharing all your words, Papa.

Thank you to our agent Jud Laghi for fighting for this book for the years (yes, years!) it took to find the right home. Speaking of which, the book you hold in your hand would not exist without the talented team at Weldon Owen, especially our fantastic, genius, un-gross-out-able editor, Mariah Bear.

And of course, no acknowledgments page worth its salt would be complete without thanks to vaccines, without which a lot of us would be dead or seriously ill. Thanks, vaccines, for being so safe and effective.

weldon**owen**

President & Publisher Roger Shaw
SVP Sales & Marketing Amy Kaneko
Associate Publisher Mariah Bear
Associate Editor Ian Cannon
Creative Director Kelly Booth
Art Director Allister Fein
Designer Scott Erwert
Illustrator Teylor Smirl
Production Director Michelle Duggan
Imaging Manager Don Hill

Weldon Owen would like to thank Brittany Bogan for editorial assistance and Kevin Broccoli of BIM for the index.

Published by
Weldon Owen International
1045 Sansome Street, Suite 100
San Francisco, CA 94111
www.weldonowen.com

ISBN 978-1-68188-381-6
10 9 8 7 6 5 4 3 2
2018 2019 2020 2021
Printed in the USA

PHOTO AND ILLUSTRATION CREDITS

All illustration by Teylor Smirl unless otherwise noted. iStock: 146, 147; B Christopher/Alamy Stock Photo: 28; Shutterstock: 22-23 (icons), 24, 29, 33, 37, 40, 42, 52, 55, 63, 75, 81, 88,89, 99, 132-133, 149, 155, 156, 97; Wissedition Ltd. & Co. KG / Alamy Stock Photo: 144-145.

Heartbreakingly, dear reader, our time together has nearly drawn to a close. But fear not—new episodes of *Sawbones* will be waiting for you weekly wherever fine podcasts are sold; maybe you'll join us there if you don't already. We hope you've found yourself entertained, enlightened, educated, edified, ejaculated and embalmed. Watch out for snake oil, get your vaccines and, as always . . .

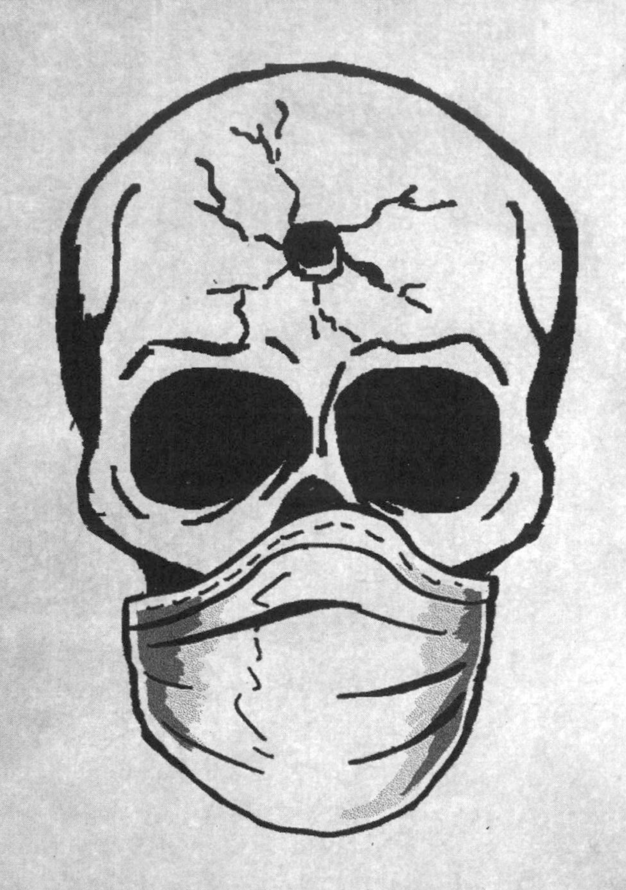

DON'T DRILL A HOLE IN YOUR HEAD!